Ethics for Biomedical Engineers

Jong Yong Abdiel Foo • Stephen J. Wilson
Andrew P. Bradley • Winston Gwee
Dennis Kwok-Wing Tam

Ethics for Biomedical Engineers

Springer

Jong Yong Abdiel Foo
Electronic and Computer Engineering Division
School of Engineering
Ngee Ann Polytechnic
Singapore, Singapore

Andrew P. Bradley
School of Information Technology and Electrical Engineering
The University of Queensland
St Lucia, QLD, Australia

Dennis Kwok-Wing Tam
Electronic and Computer Engineering Division
School of Engineering
Ngee Ann Polytechnic
Singapore, Singapore

Stephen J. Wilson
School of Information Technology and Electrical Engineering
The University of Queensland
St Lucia, QLD, Australia

Winston Gwee
Electronic and Computer Engineering Division
School of Engineering
Ngee Ann Polytechnic
Singapore, Singapore

ISBN 978-1-4899-9600-8 ISBN 978-1-4614-6913-1 (eBook)
DOI 10.1007/978-1-4614-6913-1
Springer New York Heidelberg Dordrecht London

Softcover re-print of the Hardcover 1st edition 2013

Printed on acid-free paper

Springer is part of Springer Science+Business Media (www.springer.com)

Preface

The need for engineering has moved from merely increasing productivity in the earlier days to almost all facets of life in the present world. The applications of technical knowledge and skills have also widened beyond the conventional engineering disciplines that can include the electrical, the electronic and the mechanical. Moreover, one needs to recognise that such applications often involve the marriage and/or selective adoption of principles from the aforementioned engineering disciplines. Presently, one good example would be the discipline of biomedical engineering. It is evident that biomedical engineering plays a vital role in the advances of both the medical sciences and the life sciences disciplines. Particularly, engineering principles are increasingly sought in areas such as enhancing the quality of life for patients and in the delivery of therapeutic treatments. With the proximity of biomedical engineering work to the human body, ethical practices of the biomedical engineering professionals in the workplace become just as important as those of other healthcare professionals including the medical doctors, the allied health and the nurses.

The once dogmatic belief that the study of ethics is of lesser relevance to the engineering professionals is soon becoming a dwindling past. With a number of high profile global incidents involving technological glitches, there is a growing sentiment that ethical topics need to be incorporated into engineering curriculum at the universities and colleges, as well as continual education programmes for existing engineering professionals. Likewise, this is applicable to the biomedical engineering discipline. In fact, it is more imperative for the biomedical engineering professionals to be better equipped with the understanding of acceptable practices and behaviours in their care for human lives, just as much as the other healthcare professionals. Broadly, the work involving a biomedical engineering profession can be revolved around a few major areas that include clinical engineering, medical instrumentation, implants and data mining. Therefore, it is essential for the biomedical engineering professionals and students to better appreciate the greater role this profession plays in the workplace and the responsibilities that tagged with such a role. It is hoped that through this book, it provides the necessary materials to prepare and equip the biomedical engineering professionals and students for the aforementioned purposes.

Singapore, Singapore Jong Yong Abdiel Foo

Contents

Chapter 1
Ethical Practices and Engineering

Jong Yong Abdiel Foo

Keywords Moral and social obligations • Ethical principles and theories • Best practices • Regulatory bodies • Professional societies • Education institutions • Healthcare establishments

The Need for Ethics

Traditionally, engineering has been regarded as a profession which acquires and applies scientific knowledge and technical know-how to the designing and developing of machineries, materials, devices or structures to improve the daily lives of people. In particular, the principles of engineering have been applied extensively in many sectors of the industry and society. Global recognition of engineering contributions is evident with the establishment and growth of many prominent multinational corporations like the General Electric Company and the Siemens AG that focus on engineering-related businesses. It is believed that these multinational corporations are investing most of their resources on research and development efforts to further enhance their scientific and technical capabilities. Similarly, many developed countries such as Singapore are also setting aside a substantial portion of their gross domestic product in technological research and development (The Research, Innovation and Enterprise Council 2010). While the advancement of technology has brought about many improvements and conveniences to the lives of people, it can also inflate the damages to human lives when mishaps involving technology occur. This may have to do with the general view that all necessary precautions are taken before any work is carried as illustrated in Fig. 1.1.

J.Y.A. Foo (✉)
Electronic and Computer Engineering Division, School of Engineering,
Ngee Ann Polytechnic, 535 Clementi Road, Singapore, Singapore 599489
e-mail: faj2@np.edu.sg

J.Y.A. Foo et al., *Ethics for Biomedical Engineers*,
DOI 10.1007/978-1-4614-6913-1_1, © Springer Science+Business Media New York 2013

Fig. 1.1 A typical construction site where the general assumption is that due diligence has been taken by all parties involved before any work is being carried

With the turn of the millennium, there is increasing public awareness of adverse events involving engineering failures that lead to lives being lost or those that can potentially cost lives. The key focus is not so much about the technical failure itself, but rather it is when the occurrence of such events is due to the negligence of people, especially those who could have made a difference in the outcome of the event. An example would be the collapse of the Nicoll Highway when there was an on-going construction of an underground tunnel for the mass rapid transit project in Singapore in April 2004 (The Straits Times 2004). Besides the intensive amount of the damages caused, the public's outcry was on the death of human lives and injuries that could have been avoided. More recently in the aviation industry, the Rolls-Royce Group was in the spotlight for the Trent 900 engines developed for the Airbus A380 aircrafts (BBC News 2010). Although no human lives were lost, there were questions of whether due diligence was exercised by the Rolls-Royce engineering team on the suitability of the Trent 900 engine on the Airbus A380. Obviously, the incident has also left the Rolls-Royce Group to manage the many financial, business and media implications.

In the midst of these adverse events, the job of an engineering profession is no longer just evolving around the technical know-how and development. Figure 1.2 shows the typical skillsets acquired through a conventional engineering program. From this figure, it can be seen that the area of ethics has not been evident. However, more promotion and assertion of ethical practices within the engineering profession are gradually increasing globally. Teaching ethics had been widely seen as a niche area to be taught only in humanities schools or courses. However, society as a whole is moving away from this dogmatic view because ethical practices in any profession seem to become more imperative. Professional ethics for engineering is gaining grounds to be recognised as an area to be taught in engineering-related formal

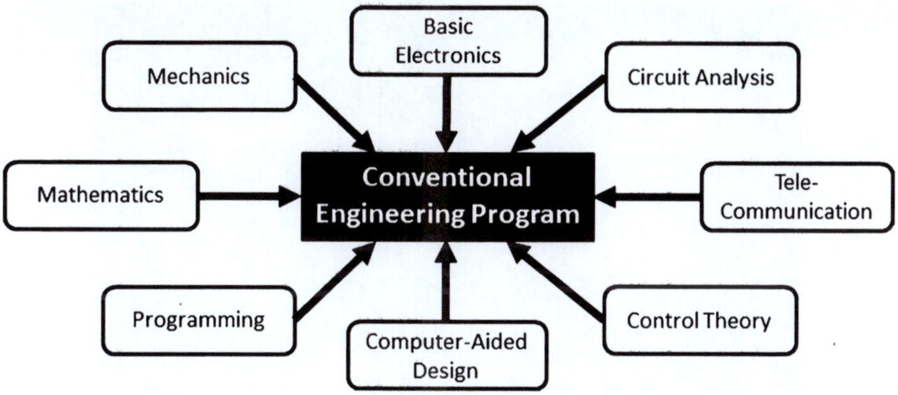

Fig. 1.2 The typical technical skills a person is expected to obtain through a conventional engineering program

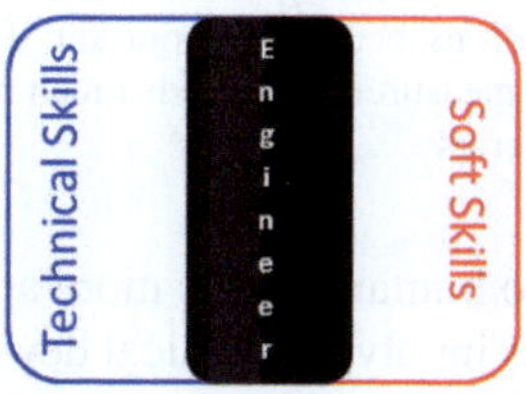

Fig. 1.3 An engineer in the twenty-first century needs to be competent not only in technical skills but also increasing the soft skills such as communication and creativity. Ethics is widely seen as part of the soft skills an engineering professional requires

education in both the universities and colleges. The word "ethics" as defined by the online Oxford Dictionary is the moral principles that govern a person's behaviour or the conducting of an activity (Oxford University Press 2011). More specifically, ethics for engineers can be defined as a field of applying a system of moral principles to the practice of engineering. It examines and sets the obligations of engineers not only to their clients but also to the engineering profession and the society as a whole. From an academic point of view, ethics for engineers is closely associated to topics such as the philosophy of science, the philosophy of engineering and the ethics of technology. In short, learning ethics may be one of the many soft skills an engineering student is required to have as illustrated in Fig. 1.3.

Presently, there are many disciplines within the engineering field and it would be impossible to adequately cover all these disciplines in this book. In view of this, the book will focus mainly on the discipline of biomedical engineering. Compared to most conventional engineering disciplines, biomedical engineering is a relatively new discipline where engineering principles are adopted in the design concepts and methodological approaches for the disciplines of medicine and biology. Biomedical engineering not only poises to be multidisciplinary but more importantly also combines the knowledge of the aforementioned disciplines to enhance healthcare diagnosis, monitoring and therapy. Due to the nature of biomedical engineering work,

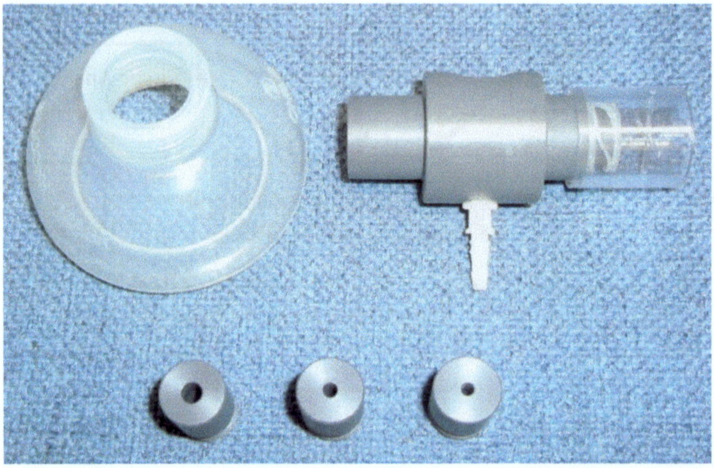

Fig. 1.4 As the work a biomedical engineering professional involves is closely related to human lives, knowing the ethical implications becomes important. The development of apparatus like those in the picture for respiratory measurements needs much more special attention as compared to apparatus developed for machineries

the impact of this discipline to human lives is more apparent and immediate per se. In other words, adverse events involving medical devices usually attract much public attention. For example, the harmful side effects of using Medtronic's Infuse Bone Graft in spinal fusion surgery were kept under wraps by the corporation. However, independent clinical studies have shown otherwise and there were many public questioning about the safety and efficacy of that product being used on patients (Carragee et al. 2011a, b). The much-drawn public's responses are understandable as this could possibly be due to the various war crimes devaluing human lives that have scarred the world. Examples in particular would be the extensive use of unconsented human subjects in medical experiments during the Nazi's Holocaust (Katz 2011; Harran et al. 2000) and the Japanese's army Unit 731 (Williams and Wallace 1989) during the World War II. Since the work of biomedical engineering does involve some experimentation on human lives directly or indirectly, the general public would also pay more attention on the development of this field. Figure 1.4 shows experimental apparatus used in a respiratory-related study.

With the proximity to human lives, a biomedical engineering professional can intervene with procedures involving medical instrumentations or devices. In other words, it is more than just performing a job of technical nature but rather there are moral and social obligations in the decisions made by a biomedical engineering professional. Understanding ethical practices can provide biomedical engineering professionals with a new set of awareness which will be required for them to manage the subtle responsibility given to them. It is also vital to recognise that the application of ethical measures will vary from people to people, even among co-workers. Therefore, the approach the biomedical engineering professional adopts is never completely technical but must continue to incorporate a wide array of moral and societal perspectives, without sacrificing sound science and good design principles (Martin et al. 2005). In a nutshell, the essence of ethics is about protecting and enhancing life. In this regard, it is then important that every biomedical engineering professional has a foundation in the topic. It is hoped that this book gives a biomedical engineering professional the tools necessary to recognise and approach ethical issues with the understanding that application of these tools may often not reach any consensus, even amongst other healthcare workers which can include fellow biomedical engineering professionals. However, the main thing one needs to understand is the context of ethics first to oneself, then how this is applicable in a specific circumstance and how that circumstance should not affect one's decision.

Ethical Principles and Ethical Theories

Decision making is part and parcel of life where different individuals adopt a variety of tools in the process of making a decision. However, it is rare for one to instantaneously identify a situation that has an ethical implication. Yet this awareness is a crucial first step and therefore recognising the moral context of a situation must precede any attempt to solve it. There are ethical principles and ethical theories available that form the foundations of ethical analysis. To begin with, the correlation between ethical principles and ethical theories needs to be established. Let us start with the latter. Ethical theory is the viewpoints from which guidance can be obtained in the formation of a decision. Each theory emphasises different points such as anticipating the possible outcome and following one's obligations to fellow human beings in order to reach an ethically correct decision. In addition, for an ethical theory to be of any use, that theory must be directed towards some common goals. Ethical principles are the common goals that each theory tries to achieve in order to be deemed as useful. The few common goals can include beneficence, least harm, respect for autonomy and justice (Beauchamp and Childress 2008; Ridley 1997). Perhaps, one of the most widely used frameworks is the Beauchamp and Childress' Four Principles. It provides a broad consideration of ethical issues especially in a medical setting which biomedical engineering is part of. The Four Principles are general guides that leave considerable room for judgement in specific cases and they are beneficence, justice, non-maleficence and respect for autonomy as shown in Fig. 1.5 (Beauchamp and Childress 2008; Walrond 2005).

First, the principle of beneficence guides the ethical theory of doing what is good. Specifically, this can mean balancing of the benefits of treatment against the risks and costs involved. The intervention of an emergency department physician in the treatment of a suicidal patient is an example of this principle. The physician acts to save the life of the suicidal patient with the belief that the patient's life is compromised and that he cannot act in his own best interest at that point of time. Second, the principle of justice requires the ethical theory to prescribe consistent actions such as distributing benefits, risks and costs fairly to all in similar positions. Cases with extenuating circumstances must contain a significant and vital difference from similar cases so as

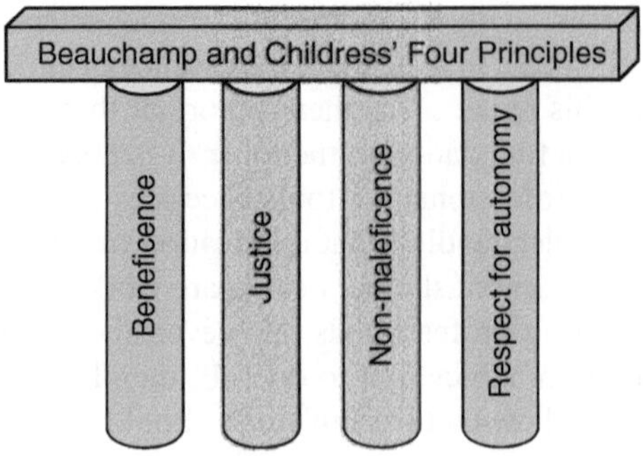

Fig. 1.5 One of the most widely used frameworks to derive ethical principles and theories is the Beauchamp and Childress' Four Principles, namely; beneficence, justice, non-maleficence and respect for autonomy. It provides a general guide to ethics issues and leave considerable room for judgement on a case-by-case basis

to that justify the inconsistent decision. For example, an ambulance is allowed to beat the traffic light if there is an emergency case where a life is endangered. Although the ambulance would normally have to obey the speed limit, it is allowed to speed off in this unique situation because it is justified under the extenuating circumstances. Third, the principle of non-maleficence directs the ethical theory to avoid the cause of harm to individuals. This is similar to the preceding principle except that the choice to do the least harm possible and/or to do harm to the fewest people should be taken. For example, a terminally ill patient may want to forego the use of life-sustaining technology because of the belief that prolonged living with a painful and debilitating condition is worse than death. In such case, only the patient alone can define what is of greater or lesser harm. Fourth, the respect for autonomy principle states that an ethical theory needs to respect the decision-making capabilities of every individual and to allow individual to make reasoned as well as informed choices. The ideology is that people should have control over their lives as much as possible because they are the only people who completely understand what they want. For example, a surrogate mother after carrying the pregnancy of another couple's child to term becomes too attached to the child and decides not to give it up to the couple. In this case, the surrogate mother has to consider not only the moral issues but also the legal implications (Beauchamp and Childress 2008; Walrond 2005; Ridley 1997).

On the other hand, ethical theories are usually based on the ethical principles that are discussed previously. However, each theory emphasises on specific aspects of an ethical dilemma and through the guidelines defined by the theory itself seeks to derive the most ethically correct resolution in each incidence. Generally, the choice of any ethical theory is dependent on the life experiences of an individual and sometimes on the circumstances the individual is faced with at that point of time (Ridley 1997). While there are a number of ethical theories in the literature, only two are covered within the context of this chapter. The selection of these two ethical theories is by no means of any significance of the two nor undermining the value of other ethical theories. The theory of deontology and utilitarian will be briefly discussed.

The theory of deontology states that people should adhere to their independent moral obligations and duties when managing an ethical dilemma. This means that for an individual to make the correct moral decision, the individual needs to clearly understand the moral duties and also the correct rules that exist to regulate those duties. Thus, when the individual follows those duties, that individual is considered to be behaving morally. As applying this theory is regardless of any person or society, an individual following this theory should produce fairly consistent decision since the decision is based on the individual's set duties. However, it is noted that the definition of duties and obligations in this theory must be determined objectively and absolutely, but not subjectively. Some have viewed this theory as dogmatic. In addition, some have argued that there seems to be no rationale or logical basis to determine the exact duties of an individual. This can then vary from individual to individual (Beauchamp and Childress 2008; Walrond 2005). Conversely, the theory of utilitarian states that an individual should make an ethical decision that yields the greatest benefit to most people. The rationale is based on predicting the consequences of an action by comparing similar predicted solutions and then using a point system to determine which choice is for the greater good. The point system provides a logical and rationale approach for each scenario and allows an individual to assess on a case-by-case basis. However, there are weaknesses with such an approach in that while knowledge and life experiences are useful in predicting outcomes, no individual can be fully certain that the prediction is flawless. It is believed that unexpected results can occur, making that individual appearing to be unethical as the decision in the end did not benefit the most number of people as predicted. Furthermore, the theory may be limited in cases where an individual needs to compare solutions on a different quality such as tangible gains like money against intangible gains like happiness (Beauchamp and Childress 2008; Walrond 2005).

It is recognised that there are variations in the definition and scope of the ethical theories and ethical principles to that being described in the text. Thus, in this aspect, the coverage herein is not all encompassing but it is hoped that sufficient basis is provided to understand the vast spectrum of the two in this literature.

Code of Conduct and Best Practices

In biomedical research, there are international guidelines such as the Nuremberg Code and the Declaration of Helsinki developed to regulate experimentation on human subjects. The former is a set of research ethics principles for human experimentation set as a result of the subsequent Nuremberg trials at the end of the Second World War, while the latter is developed for the medical community by the World Medical Association which is widely regarded as the cornerstone document of human research ethics. At this point in time, it may not be a legally binding in international law, but instead it draws its authority from the degree to which it has codified in, or influenced, national or regional legislation and regulations (Walrond 2005; Bosnjak 2001). However, this is less obvious in the engineering field including BME until the turn of the last millennium. This may be due to the dogmatic association of engineering with technology or machinery, and it being perceived as a less humanistic profession.

With increasing public awareness and attention on the adverse events involving medical devices, multi-faceted approaches have been adopted in order to accelerate the permutation of ethical knowledge and practices in the biomedical engineering field. There are also strong evidence in the literature that supports the vital role of ethics practices to the integrity of science (Monzon and Monzon-Wyngaard 2009; Ross and Athanassoulis 2010). On one hand, it would be hard to find any government or academia opposing to the responsible conduct in the biomedical engineering field, but the implementation of such practices somehow seems lacking. However, consensus and commitment to the "best practices" can disappear when it comes to promoting instructions in responsible conduct. It is also known that human experimental science has been around for at least a few centuries and it seems that the aforementioned difficulties have yet to be adequately addressed. Hence, this can suggest that there are simply unfathomable challenges to be answered before "best practices" can be widely adopted. Essentially, this may include the difficulty to quantify the goals, the methods for achieving the goals and the willingness of organisations or individuals to pay for or commit to such programmes (Kalichman 2007).

In the recent years, the difficulties to implement "best practices" seem to become less apparent with greater public outcry and governmental commitment. In particular, there are few common approaches adapted to overcome the hurdle or extrinsic factors which can include through the governance of regulatory bodies, the recognition from professional societies, the inclusion in syllabus for education institutions and the participation of healthcare establishments. The bottom line for all these approaches is the translation of "best practices" in terms of patient safety and to enhance the use of technology for the welfare of patients or human lives as a whole. It is worth noting that there are both marketplace and regulatory requirements to be fulfilled. Broadly speaking, there are at least two groups of biomedical engineering professionals that need to be equipped with the "best practices" of the field; existing personals in the workplace including personals to be cross-trained, and students enrolled in a biomedical engineering-related programme. Similar to other professions or fields, both groups may probably require different levels of immersion to comprehend the ethical expectations.

The term "best practices" may be commonly used but it is also subjected to different interpretations. Moreover, it will continue to defy its dogmatic definition as its requirements will evolve partly due to responses from the increasing concerns about highly publicised adverse events as discussed previously (Carragee et al. 2011a, b). While in a typical training or definition of ethical practices, topic areas such as conflict of interest, confidentiality and responsible conduct are easily accepted and to some extent practised in the workplace. Conversely, topic areas such as policies for handling misconduct and the costs of whistleblowing may appear to be lacking or being downplayed. Above all, defining the scope of ethical practices may just be one small aspect of the whole equation. The crux of the matter is not about encompassing a wide-ranging mix of ethical knowledge and rules following. It is about the intrinsic values of an individual that is having good characters, exhibiting good ethical judgement and acting with integrity and responsibility. Moreover, the extent to which ethical practices should cover is further highlighted by many publications in the literature which report a diverse and sometimes conflicting array of topic areas that can be classified into domains of knowledge, skills, attitudes and behaviour. All these topic areas are to be valued, but that does not mean that they need to be, and realistically can be, the purpose of understanding ethical practices (Kalichman 2007; Walrond 2005).

At a quick glance, it looks hopeless to reconcile the diverse topic areas for "best practices" in the biomedical engineering field, but it is worth taking a closer look to see whether there is common ground for a clear and manageable inclusion. To begin, rather than focusing on specific topic areas or objectives for either group of biomedical engineering professionals, it may be more beneficial to address the long-term impact on the biomedical engineering field. Probably, the two most obvious outcomes are less adverse events involving medical devices and more responsible conduct. Simply, it is about promoting a common understanding and practice of accepted standards of care for human lives. This scenario is not only desirable to the patients but also it can promote an environment that is well supported and effective amongst the biomedical engineering professionals. Figure 1.6 presents a possible scheme for the desired outcomes and factors that can have an impact on these outcomes.

Decreasing adverse events involving medical devices and increasing the likelihood of responsible conduct presumably depends on much more than a framework of "best practices" whether constitutionalised or being taught in formal education. As mentioned, the inculcation of "best practices" can be broadly defined as two major groups; existing personals and biomedical engineering students. Between the two groups, the latter would generally be easier to impress on them the importance of ethical practices through formal education. However, it is noted that intrinsic factors such as upbringing and family setting are vital considerations for individual student's behaviour. For the former group, there is additional variety of intrinsic factors as compared to the latter group. First and foremost, an individual would bring a personal background to the workplace. Individuals vary in life experiences that have shaped their moral disposition and character. Moreover, an individual's conduct also depends on the strength of that individual on moral reasoning and ethical decision-making skills. It is worth noting that the development of these skills is frequently perceived as an outcome of formal education or an extrinsic factor. However, this may be just more related to the critical thinking skills developed by that individual during any previous education training. The issue with ethical practices is not that individuals lack the necessary reasoning skills, but perhaps that they lack either access to the necessary information or recognition of the need to apply their skills to the ethical dimensions of their professional practice. Whether an individual adopts "best practices" in their workplace depends not only on their character and knowledge but also on the environment the individual works in. There are a

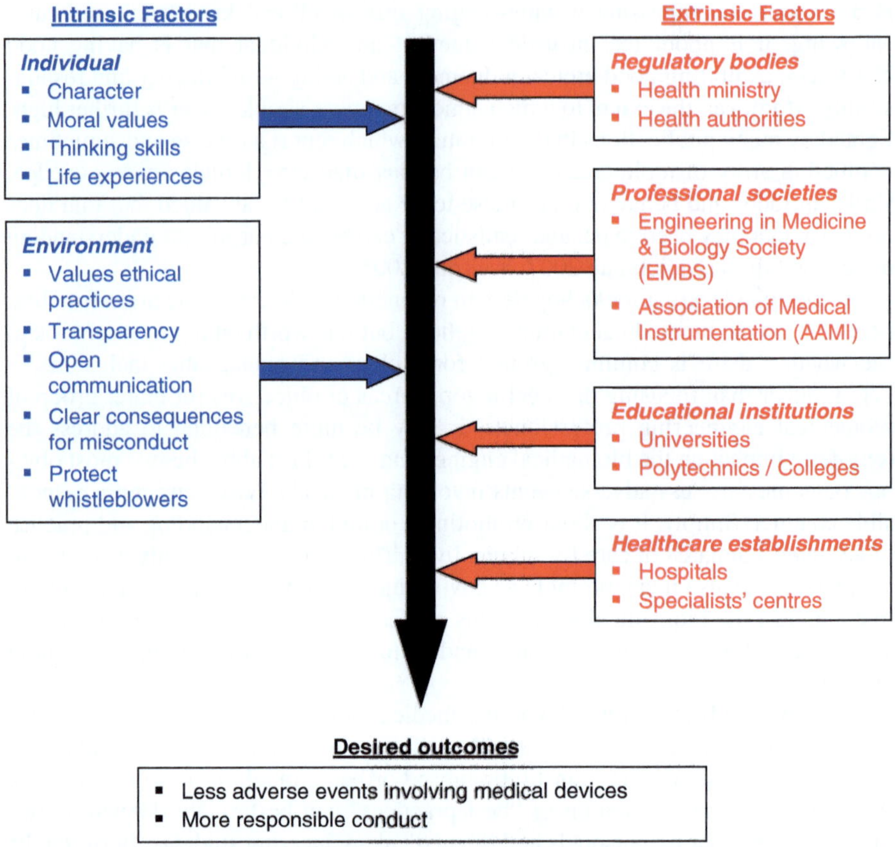

Fig. 1.6 A relationship between intrinsic factors of the workplace (moral values, life experiences and environment), extrinsic factors (regulatory bodies, professional societies, educational institutions and healthcare establishments*) and actual "best practices". *Refer to section "Contributions of Education Institutions" for the explanation of including the healthcare establishments as an extrinsic factor rather than otherwise

number of environmental factors that would impact an individual to act responsibly or police when required, but some of these factors are likely to be more important. In order to determine the importance of each factor, the following thoughts can be evaluated:

- Does the workplace environment place a strong emphasis on ethical practices?
- Is there regular training or updates on the value of ethical practices?
- Are there avenues available for individual to voice out concerning misconduct?
- Is there open communication to foster transparency in defining "best practices"?
- Are the consequences of misconduct clear and appropriate?
- Does the environment protect responsible whistleblowing and potential ethical issues?

Based on the materials covered thus far, it may seem that whether an individual acts ethically in the profession depends greatly on that individual's character and the work environment the individual is placed in. Thus, it appears to have little or no benefit in emphasising the value through the education system or being imposed on. However, this is far from the truth where the contributions of the regulatory bodies, the professional societies, the educational institutions and the healthcare establishments must not be undermined. Through these channels, biomedical engineering professionals are kept abreast of their moral, professional as well as legal obligations. In the unfortunate occurrence of misconduct, act of ignorance can then not be used as plead for legal mitigation. Next, the roles and efforts for each of the extrinsic factors are discussed in greater details.

Contributions of Regulatory Bodies

Policymakers have a great responsibility to influence most, if not all, the outcomes and well-being of the nation. Likewise, ethical practices in the biomedical field need to be constituted to have the fullest impact and minimising mischievous acts from occurring. Simply, country ownership is the surest way to permeate "best practices" amongst any professions. Essentially, there are at least four key steps in making ethical practices throughout the country a reality. First, policymakers must plan with a clear development vision and a detailed roadmap for realising it. For this to be achievable, development partners including countries and/or multinational corporations with the specific domain knowledge need to be identified. Policymakers should also be open to ideas and seek to tailor-proven practices to the actual circumstances of the country. Once a plan is in place, these partners are still needed to support that plan and advise when necessary to ensure it thrives. Second, policymakers must ensure that there are enough resources to execute the plan. If resources are limited, careful prioritisation is crucial. Policymakers should have in mind contingency plans that focus on the more pressing priorities rather than the broader and more ambitious portion of the plan. Third, policymakers need to take measures to implement the plan. This may be the most crucial part when the country must be fully committed and engaged. Probably, the most efficient and sustainable approach in implementing the plan is to ride on existing structures and/or capacities within the country rather than replacing with parallel efforts. Fourth, the plan needs to be monitored and evaluated. At this point, the partners can provide valuable inputs to clear outcome targets and performance tracking as defined at the outset.

In the biomedical field, an example of a country where good level of commitments to ethical practices is observed is in the island state of Singapore. Besides its health ministry that drafts policies and guidelines related to ethical practices, a statutory board called Health Science Authority has been set up to regulate the field. The health ministry has guidelines like relating to ethics (encompassing ethics committee and research ethics), professional practices and for private healthcare institutions (Ministry of Health, Singapore 2007). On the other hand, the Health Science Authority provides frameworks that not only enable consumers to have safe and

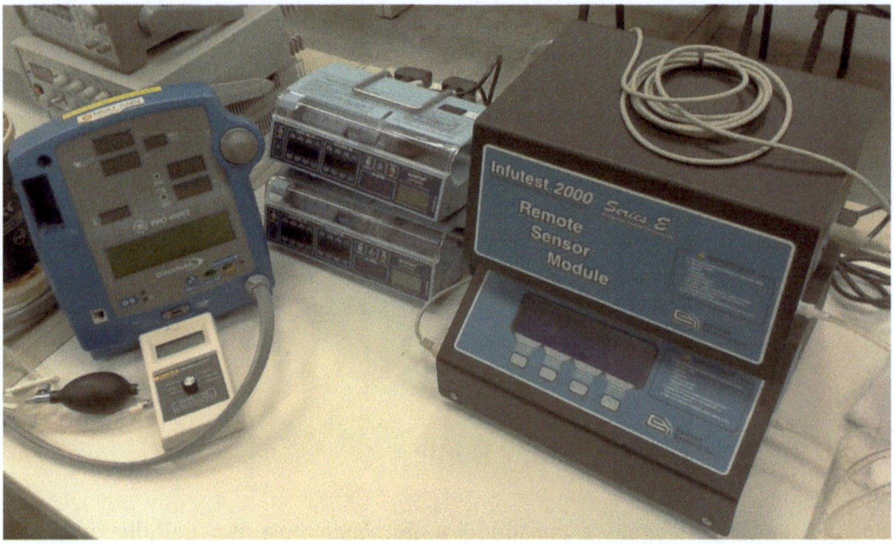

Fig. 1.7 The Health Science Authority of Singapore is the equivalent of the Food and Drug Administration in the USA where its key mandate is to regulate policies and guidelines pertaining to all medical devices. The figure shows some examples of the medical equipment and devices (from *left*: non-invasive blood pressure monitor, syringe, syringe pump and infusion device analyser) that are regulated

timely access to health products but also take into account industry needs for greater transparency and flexibility. The regulatory body is also recognised by international entities such as the World Health Organisation. In addition, it collaborates with global similar agencies like the United States' Food and Drug Administration, Health Canada, Swiss Medic, the Australian Therapeutic Goods Administration and the Chinese State Food and Drug Administration, to tackle pressing issues in the regulatory sciences. In addition, the Health Science Authority manages the island-wide reporting of medical adverse events including those involving medical devices (Health Sciences Authority 2011). Figure 1.7 shows medical equipment and devices that are commonly used in the hospitals. With the combination like the health ministry with the Health Science Authority, cascading the mandatory guidelines and ethical practices would then be more manageable. It is worth noting that many other countries that include the USA, Switzerland and Australia have established their own unique frameworks in ensuring ethical practices at their national level.

Contributions of Professional Societies

It is believed that the topic of ethics is not usually taught in most traditional engineering program, hence there is a gap in the knowledge for existing engineering professionals including those that are hired as a biomedical engineering professional (Monzon 1999). One possible way is for these professionals to enroll in bridging courses that cover this gap of knowledge. However, this may not be feasible due to the cost implications and time commitments that both the employers and individuals may be reluctant to invest. Realistically, it must also be recognised that for any biomedical engineering professionals whose engineering education had been oriented towards inert materials which lack usual or anticipated action from a chemical or biological point of view, problems will arise when they need to deal with living material either directly or indirectly. Classical tools that they are familiar with were meant to work on substances that do not show a pre-established order as the living organism does. Thus, professional societies can play a vital role in a few aspects to assist existing biomedical engineering professionals to be updated in the field.

Professional biomedical engineering societies such as the Engineering in Medicine and Biology Society of the Institute of Electrical and Electronics Engineers have their codes of ethics (IEEE Engineering in Medicine and Biology Society 2011), and the Association for the Advancement of Medical Instrumentation has their standards and recommended practices (The Association for the Advancement of Medical Instrumentation 2011). These codes or standards are being derived for the purpose of educating their members as well as the biomedical engineering professionals in understanding the moral obligations and expectations of the trade. For example, the code of ethics from the Engineering in Medicine and Biology Society emphasises on a few aspects; respecting human dignity and privacy, safeguarding of confidential information, preserving a healthy working environment, avoiding conflict of interest as well as good laboratory practices and good clinical practices. These aspects may be general and simple, but it is evident that the Engineering in Medicine and Biology Society is committed to realign to "best practices" in the biomedical engineering field. In addition, both the Engineering in Medicine and Biology Society and the Association for the Advancement of Medical Instrumentation are publishing updates for best practices in the biomedical engineering field regularly through their general interest journals, the Biomedical Instrumentation and Technology and the IEEE Pulse. Particularly, the Biomedical Instrumentation and Technology has a section that publishes the best practices in clinical settings in its regular issues. One of such papers focuses on an effective approach to schedule equipment replacement in clinical environment (Williams 2011). This is essential as there is a need to balance between providing the best available equipment to patient and maintaining efficiency use of equipment. It is also interesting to note that there are international conferences that focus on ethics for biomedical engineering professions including the International Conference on Ethical Issues in Biomedical Engineering and a track on ethics in the well-established Annual International Conference of the Engineering in Medicine and Biology Society as illustrated in Fig. 1.8. From time to time, it is known that professional societies also organise forums where the specific areas of the ethical topic are discussed. However, this type of forums is not as well publicised and is generally meant for more localised audiences.

Thus, it is clear from the efforts of the various professional societies that there are a few key aspects that a biomedical engineering professional should be aware of,

Fig. 1.8 Events organised by the professional societies like international conferences are good platforms for exchange of ideas or cutting edge knowledge in the field

namely; respect for patients and/or human subjects, handling of confidential information, research conduct and basic profession conduct. However, it is worth noting that the onus in this case falls heavily on the initiative of individuals in wanting to keep abreast of the development of the field.

Contributions of Education Institutions

As discussed much earlier, not only existing biomedical engineering professionals need to be adequately fluent in ethical practices but also students currently enrolled in a formal education program-related biomedical engineering. For the former, the employers and biomedical engineering professionals themselves need to come to the realisation of the significance of ethical knowledge and practices. Besides learning through the avenues of professional societies, the former can also enrol in short bridging courses in the universities or colleges to be formally equipped in the arena. There are potentially two modes of learning for these adult learners; online courses or evening classes. An example of the online courses is that of the online Master's course from the Purdue College of Engineering with a concentration in biomedical engineering where the ethics module is part of the curriculum (Purdue University 2011). It is also possible for the biomedical engineering professionals to enrol in evening classes like the degree program in biomedical engineering by the Singapore Institute of Management which has a module in biomedical ethics (SIM University 2009–2011). However, the same point of self-initiative to such courses needs to be reiterated again. Regardless of the motivation of individuals, the educational

Fig. 1.9 Educational institutions are now expected not only to impart academic knowledge and train technical skills to their enrolled students but also to increase emphasis on moral systems and ethical values

institutions form a strong pillar in broadcasting the heart of ethical practices to the existing biomedical engineering professionals, current students and beyond. Figure 1.9 shows the St. Lucia campus of the University of Queensland, where like many other universities and colleges the students are expecting more than just academic impartation of knowledge from the educational institutions.

For the current students enrolled in fulltime programs in either universities or colleges, it would take a totally different view on ethical practices. It may also be perturbing to them why such emphasis is placed on relearning moral values when this was taught to them at a much younger age. Nevertheless, typical objectives of the biomedical engineering ethics module should include:

- Understanding the ethical issues in biomedical engineering practice.
- Formulate arguments on ethical issues.
- Be equipped with tools to make well-informed decisions about ethical issues
- Evaluate the morality of choices and decisions.
- Manage multiple points of view in decision-making process.
- Assess the impact of biomedical engineering solutions on individuals, communities and the environment.

To equip the students with these skills, some of their common lectures can consist of fundamentals of ethics, professionalism in engineering, conflict of interest, intellectual property, ethical regulations and ethics in engineering design. These lectures component can vary from each institution. However, an important component that is generally adopted to facilitate the learning experiences of the students is the use of case studies. It is believed through case studies that the students better develop the aforementioned objectives.

For many educational institutions, the mandate of education also includes conducting research to acquire new knowledge to further enhance the capabilities of the staffs as well as imparting it to the students. For example, a research centre in Singapore that focuses on ethics has been established in the National University of Singapore, the Centre for Biomedical Ethics. The centre has the ethos to become a major location for teaching and research in South East Asia and its main focus is on the ethical values in an Asian context. Besides forging interdisciplinary collaborations with other faculties in the university, the centre also initiates research efforts with the healthcare sectors and the research communities both locally and internationally. Besides working with the ethics governance and advisory bodies in Singapore, the centre also enhances public understanding of ethical issues in biomedicine (National University of Singapore 2001–2008). Hence, the twofold contributions of educational institutions with regard to ethical practices cannot be undermined.

Contributions of Healthcare Establishments

It is recognised that the healthcare establishments are one of the major employers in the biomedical engineering field. Thus, it may seem strange that it is considered as an extrinsic factor rather than intrinsic factor (that is, workplace environment). However, it is worth noting that the public healthcare establishments are the ones that generally set the precedence in the standards in the field. Thereafter, their private counterparts and industry would then follow suit. It is from this point of view that the contributions from the healthcare establishments as an extrinsic factor are being considered. To begin with, the healthcare establishments are no strangers to ethical regulations as health professionals including the clinicians are expected to adhere to the Hippocratic Oath that they swear to practice medicine ethically. Besides their daily practices, healthcare professionals wanting to conduct clinical trials or research within the establishment are under strict regulations. They are required to submit properly documented paperwork detailing the purposes, expected outcomes, possible detriment effects, recommended remedial actions and benefits to the community for their proposal. A committee, independent from the said work, will be formally designated to approve, monitor and review the proposal with the aim to protect the rights and welfare of the patients or human subjects in spite of potential scientific and/or medical benefits. This committee can be known as the institutional review board, ethics committee or ethical review board. Generally, the composition of an institutional review board follows the guidelines from the United States' Food and Drug Administration (U.S. Food and Drug Administration 2011) that includes:

- Each institutional review board shall have at least five members, with varying backgrounds (such as race, gender, culture and sensitivity to community issues). If the work involves vulnerable subjects such as children, prisoners, pregnant women, handicapped or mentally disabled persons, inclusion of individual(s) who are knowledgeable about and experienced in working with those subjects is needed.
- Each institutional review board shall not consist entirely of men or women and shall not consist entirely of members of one profession.
- Each institutional review board shall include at least one member whose primary concerns are in the scientific area and at least one member whose primary concerns are in non-scientific areas.
- Each institutional review board shall include at least one member who is not affiliated with the institution and in the immediate family of a person affiliated with the institution. Generally, such institutional review board member(s) are known as the community member(s).
- No institutional review board member shall participate in the review of member's own project(s).
- Any institutional review board may include consultants in their discussions to meet requirements for expertise or diversity, but only actual institutional review board members can vote.

On top of the Hippocratic Oath and the institutional review board, healthcare establishments may have additional codes of ethical standards for specific patient groups like the Aboriginals and Torres Strait Islanders in Australia (National Health and Medical Research Council, Commonwealth of Australia 2011). Since biomedical engineering equipment is generally applied onto patients or human subjects, there are also tight compliance requirements for all biomedical engineering equipment. Broadly, the international standard of the International Electro-technical Commission 60601-1 is used to outline the basic safety and essential performance of any biomedical engineering electrical equipment. While there may be compliance requirements for the biomedical engineering equipment, biomedical engineering professionals may not be as competent in managing ethical dilemma as their other healthcare counterparts. This may be more pronounced in biomedical engineering professionals who have undergone more traditionally engineering education and when they are in need to develop engineering devices such as those in Fig. 1.10 to be used in the healthcare industry. However, this can be put up for debates. In addition, it can also be argued that biomedical engineering professionals whom have worked in or with a healthcare establishment may differ in their understanding of ethical practices when compared with those who have not. The important point to take note here is that previous expositions to ethical practices in any healthcare establishment would be helpful for a biomedical engineering professional in the ethical outlook and the management of such issues. Thus, the ethical contributions of any healthcare establishment back to its own arena as well as other biomedical engineering arenas like research institutes and industry are significant in this regard.

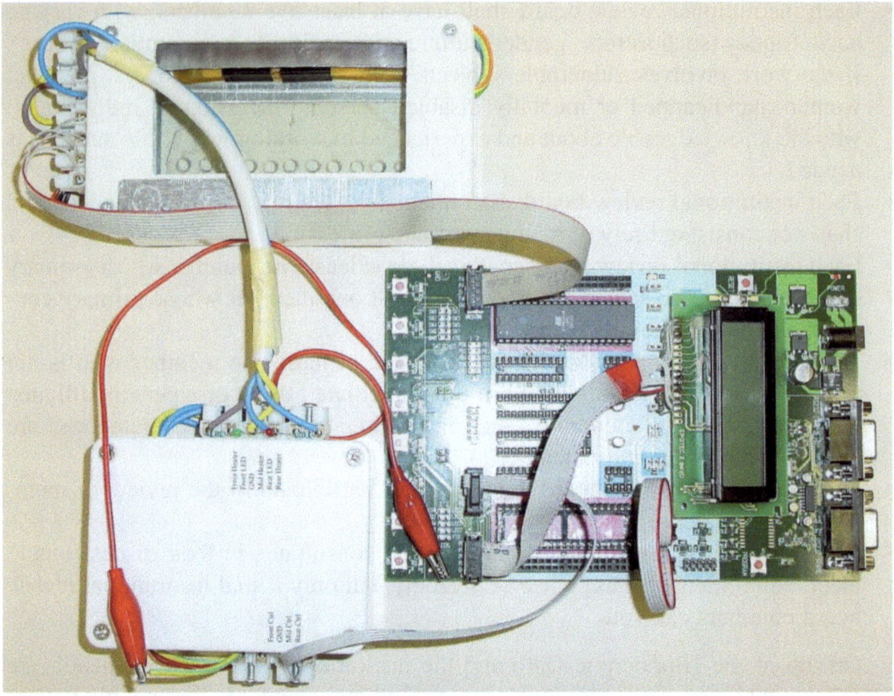

Fig. 1.10 In a healthcare establishment, a biomedical engineering professional needs to be mindful of the institutional review board approval and the International Electro-technical Commission 60601-1 compliance when developing the firmware and prototyping of a device to be used for research or other purposes within the establishment

Ethics and Biomedical Engineering

While the decision made in an ethical dilemma is mainly depended on the moral values of an individual, the impact of one's life experiences and workplace environment may have confounding effects on the actual outcome. Particularly, if one had working experiences with an organisation where unorthodox practices were seen as the norms, one may still bring these practices to one's future employer. Hence, continuous efforts to create organisational awareness become a critical element to instil the "best practices" mentality in all the employees. As discussed previously, the contributions of the various extrinsic factors like regulatory bodies, professional societies, educational institutions and healthcare establishments must not be undermined as well. Presently, there are possibly a number of work areas the biomedical engineering professionals can be involved with that can include:

- Work in a clinical setting or clinical engineering
- Develop and service medical instrumentations
- Relevance of medical ethics to biomedical engineering

- Research and develop biomaterials and implants
- Provide statistical analyses and data mining

The next few chapters of this book will focus on the said work areas of biomedical engineering including the nature of the work and how ethical practices are essential to each area accordingly. It is acknowledged that these work areas are not the exhaustive list of which a biomedical engineering professional can be involved with.

References

Beauchamp TL, Childress JF (2008) Principles of biomedical ethics, 6th edn. Oxford University Press, New York, NY

Bosnjak S (2001) The Declaration of Helsinki—the cornerstone of research ethics. Arch Oncol 9(3):179–184

Carragee EJ, Ghanayem AJ, Weiner BK, Rothman DJ, Bono CM (2011a) A challenge to integrity in spine publications: years of living dangerously with the promotion of bone growth factors. Spine J 11(6):463–468

Carragee EJ, Hurwitz EL, Weiner BK (2011b) A critical review of recombinant human bone morphogenetic protein-2 trials in spinal surgery: emerging safety concerns and lessons learned. Spine J 11(6):471–491

Harran M, Roth J, Kuntz D, Lemmons R, Michael RA, Rickus K, Aretha D (2000) The Holocaust Chronicles: a history in words and pictures. Publications International, Lincolnwood, IL

Health Sciences Authority (HSA) (2011) HSA's corporate profile [Internet] [Cited 20 Sep 2011]. http://www.hsa.gov.sg/publish/hsaportal/en/about_us/.

IEEE Engineering in Medicine and Biology Society (EMBS) (2011) IEEE EMBS code of ethics [Internet] [Cited 29 Sep 2011]. http://www.embs.org/docs/.

Kalichman MW (2007) Responding to challenges in educating for the responsible conduct of research. Acad Med 82(9):870–875

Katz E (2011) The Nazi engineers: reflections on technological ethics in hell. Sci Eng Ethics 17(3):571–582

Martin T, Rayne K, Kemp NJ, Hart J, Diller KR (2005) Teaching for adaptive expertise in biomedical engineering ethics. Sci Eng Ethics 11(2):257–276

National Health and Medical Research Council, Commonwealth of Australia (2011) Values and ethics—guidelines for ethical conduct in aboriginal and Torres Strait Islander health research [Internet] [Cited 6 Oct 2011]. http://www.nhmrc.gov.au/guidelines/publications/e52

Ministry of Health (MOH), Singapore (2007) MOH's publications overview [Internet] [Cited 20 Sep 2011]. http://www.moh.gov.sg/mohcorp/publications.aspx

Monzon JE (1999) Teaching ethical issues in biomedical engineering. Int J Eng Educat 15(4):276–281

Monzon JE, Monzon-Wyngaard A (2009) Ethics and biomedical engineering education: the continual defiance. In: Annual international conference of the IEEE engineering in medicine and biology society 2009 (EMBC 2009), 2011–2014.

National University of Singapore (NUS) (2001–2008) NUS Centre for biomedical ethics [Internet] [Cited 6 Oct 2011]. http://cbme.nus.edu.sg/.

BBC News (2010) Rolls-Royce 'makes progress' in A380 engine probe [Internet] [Cited 24 Aug 2011]. http://www.bbc.co.uk/news/11709179.

Oxford University Press (2011) Oxford Dictionaries Online [Internet] [Cited 24 Aug 2011]. http://oxforddictionaries.com/.

Williams P, Wallace D (1989) Unit 731: Japan's secret biological warfare in World War II, 1st edn. Free Press, New York, NY

Purdue University (2011) Purdue engineering professional education [Internet] [Cited 1 Oct 2011]. https://engineering.purdue.edu/ProEd/credit/bme.

Ridley A (1997) Beginning bioethics: A text with integrated readings. Bedford/St. Martin's; 1st edn.

Ross A, Athanassoulis N (2010) The social nature of engineering and its implications for risk taking. Sci Eng Ethics 16(1):147–168

The Association for the Advancement of Medical Instrumentation (AAMI) (2011) AAMI standards [Internet] [Cited 29 Sep 2011]. http://www.aami.org/standards/index.html.

The Research, Innovation and Enterprise Council (RIEC) (2010) Government commits S$16.1 billion to support research, innovation and enterprise for the next 5 years and seeks ways to solve complex national challenges with R&D. 4th RIEC Press Release, 17 Sep 2010.

The Straits Times (2004) MRT worksite collapse wrecks Nicoll Highway. 21 April 2004:H1–H3.

U.S. Food and Drug Administration (2011) CFR—Code of federal regulations Title 21 [Internet] [Cited 6 Oct 2011]. http://www.accessdata.fda.gov

SIM University (2009–2011) SIM University BSc Biomedical Engineering [Internet] [Cited 1 Oct 2011]. http://www.sim.edu.sg/learn-sim/pages/part-time-programmes.aspx.

Walrond ER (2005) Ethical practice in everyday health care. University of the West Indies Press, Kingston, Jamaica

Williams JS (2011) Right-sizing and replacing the right equipment at the right time. Biomed Instrum Technol 45(3):214–218

Chapter 2
Ethics and Biomedical Engineering Practice and Research: Origins of Principles and Consent

Stephen J. Wilson and Jong Yong Abdiel Foo

Keywords Regulatory approval • Clinical trial • Classification of medical devices • Beneficial and non-beneficial studies • Patient risk • Institutional review board • Ethics committee • Informed consent

Introduction

Modern biomedical engineering practice relies heavily on the evidence base developed from trials, evaluations and audit studies of biomedical diagnostics and therapies. It is rare for a modern-day professional not to be exposed to such research and most accept the conduct of such an activity core to the profession. The basis of how we conduct studies both with and without human subject involvement is a well-developed and developing area. Principally, how we structure and conduct any study in this field is subject to ethical review. The subject of ethics arises from the philosophical basis of how we treat each other or, more generally, how we determine right from wrong behaviour.

Here, we propose to discuss the origins of ethical principles in research conduct with a specific reference to the biomedical engineer who straddles the worlds of basic and applied science and the sometimes imprecise world of the clinician.

S.J. Wilson (✉)
School of Information Technology and Electrical Engineering, The University of Queensland, St Lucia, Brisbane, QLD 4072, Australia
e-mail: wilson@itee.uq.edu.au

J.Y.A. Foo
Electronic and Computer Engineering Division, School of Engineering, Ngee Ann Polytechnic, 535 Clementi Road, Singapore, Singapore 599489

J.Y.A. Foo et al., *Ethics for Biomedical Engineers*,
DOI 10.1007/978-1-4614-6913-1_2, © Springer Science+Business Media New York 2013

Technology and Quality of Life

It is becoming a norm where the provision of healthcare and medicinal treatments has gone more technologically advanced (Free et al. 2010). Over the past few decades, the use of computers and electronics has radically transformed everything in the medical field from home-based healthcare to surgery and rehabilitation to implants. The applications of medical technology are helping people to live longer and/or better assist them in their daily routines. Research facilitates in healthcare establishments, educational institutions, research organisations and industrial companies are continually developing a spate of medical technologies for a variety of medical conditions such as diseases, cancers and disabilities. The purpose is to collectively help medical staffs perform their tasks more efficiently and for people to live a better quality lives and more independently.

Without a doubt, the application of technology in the medical field may be obvious to many but the essential forefront step before such technologies can be rolled out is somewhat less prominent and sometimes rather unknown. The entire process of bringing a new technological approach to the medical field is pivoted on the need and success of clinical trials as well as the subsequent mandatory approvals from the local regulatory bodies. As for enhancements made to existing medical devices that include medical equipment, the need for clinical trials and regulatory approval may vary on a case-by-case basis. It must be acknowledged that the benefits of continual enhancements made to the medical devices have become more evident and proclaimed through the increased uses of mass media. Figure 2.1 shows the prototype built for a preclinical trial on an enhancement to be made on a finger probe of an existing pulse oximeter.

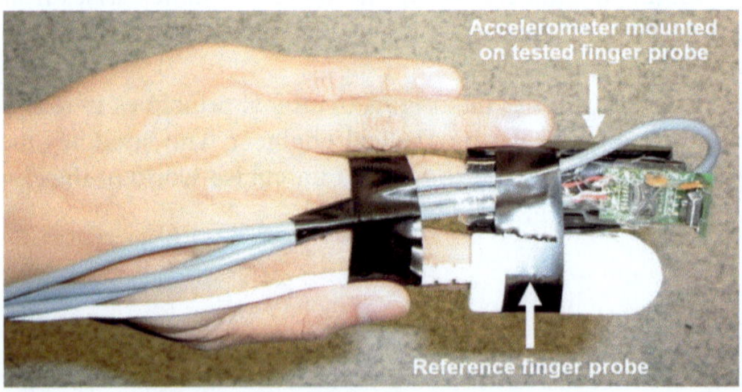

Fig. 2.1 The use of technology has improved the quality of physiological measurements and eventually the diagnosis and well-being of patients. The availability of new technological approach to the public is dependent on the positive outcomes from clinical trials

Invention and Its Necessities

Although the motivation to invent and research on new technologies for the medical field maybe clear, it must be recognised that the overheads and associated expenses for such activities can be demanding. Public-funded organisations are now working more closely with the industry to bringing newly developed medical devices to the general community. Typically, larger medical device companies develop successive iterations of existing devices while most new device categories are developed by venture-backed start-up companies (Kaplan et al. 2004). The general motivation to invent arises when there is an unmet clinical challenge. Figure 2.2 shows the process (though not exhaustive) of commercialising of a conceptual idea for a new medical device. It is worth noting that the process for enhancement(s) to be made on an existing device is often simpler. The whole conceptualisation process starts with either a medical staff or an engineering researcher conceives a solution for it which may be in a form of a device or method. The initial stage would usually see the formation of partnership between the two (that is, the medical staff and the engineering researcher). The next stage includes building a preliminary device prototype and securing some forms of research funding or vice versa. In some cases, a patent application process is initiated when deemed necessary by the partnership. Once the research fund application is successful, a team is formed where research

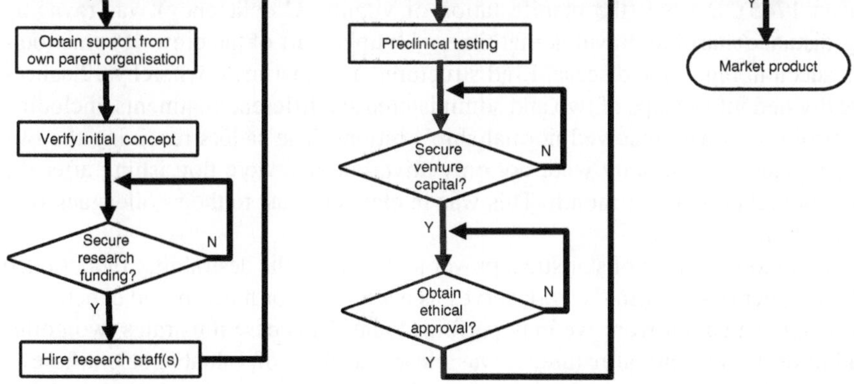

Fig. 2.2 A flowchart detailing the process of bringing a conceptualised idea to commercialisation of a new medical device. It can be seen that the process can be tedious and the required resources are usually substantial

staff(s) can be hired from the research funding to bring the concept through a "design–build–test–redesign" cycle.

Often, the research funds are also sufficient for the initial bench testing of the prototype and possibly for animal testing. Beyond that, additional funds will be needed to bring the invention through the preclinical stage which can come in the form of venture-backed start-up companies or from angel investors. At this point, intellectual property such as the technical know-how can also be licensed out to an existing or start-up company for further development. The preclinical stage can take up to 3 years or more and may consume as much as millions of dollars before the prototype is ready for clinical testing. These capital requirements usually exceed the means of most angel syndicates and only venture capital firms in the form of equity financing can keep the process going. Depending on the complexity of the device and its intended applications, the entire process from conceptualisation to regulatory approval may take up to 10 years and tens of millions of dollars can be easily spent (Aaron et al. 2004). In order to have a useful invention for any given application, it is important for one to understand and meet the need(s) for its intended usage through research and experimentations. From this perspective, the possible ethical issues a biomedical engineer may be involved with throughout the entire conceptualisation process can be quite widespread. Before we can appreciate the implicit ethical issues, it may be worthwhile to first understand why and how biomedical research was originated.

Origins of Biomedical Research

Biomedical research in a Western context, by many accounts, may trace its origins to experiments conducted in 1747 aboard the H.M.S. Salisbury where the ship's surgeon, Dr. James Lind, conducted the first experiment for the treatment of scurvy (Anson 1745). Scurvy (the manifestation of vitamin C deficiency) was ravaging long-distance maritime travel at that time with up to half of the crew of some journeys succumbing to the disease. Lind structured an experiment whereby 12 sailors were divided into groups of two and administered six different treatments, including a control group who received normal ship's rations. The sailors receiving the two oranges and a lemon daily were not only alive but also were flourishing after the experimental period of 1 month. This was in clear contrast to their colleagues who were declining rapidly.

If we ignore the lack of statistical power and perhaps the desirability of a blinded and crossover design, Lind's work serves as a template for much of the experimentation and evaluation work we may perform today. This case illustrates two points in addition to its seminal nature. It was clearly a piece of valuable scientific evidence that leads to the eventual adoption of citrus rations for the entire naval service, thus saving thousands of lives and immeasurable suffering, but also brought into focus the issue of human experimental subject autonomy.

The selection of subjects for this experiment was not voluntary. The benefits were widespread and the authority of the surgeon was not questioned. The application of authority in this context is known as paternalism and as such has been the model for most biomedical experimentation using human subject since that time until the 1930s. Such decision making or mode of thinking addresses what you "consider" is best for the subject, not what may actually be in their best interests or desires.

The concept of empowering the subject or patient is one of the first principles in the consideration of what may be classed as modern ethical research. The reason for empowerment may not be simply seen as a human right, but more as a practical tool whereby the fears, anxieties and irrational perceptions can be allayed for the benefit of both the patient and the research. The involvement of a distressed participant is clearly deleterious to any proposed experimental trial.

Clinical Trials and Its Importance

The approach to conduct experiments on human subjects to answer a medical question like what Dr. James Lind did on the ship, H.M.S. Salisbury, in 1747, is now commonly termed as clinical trial. Clinical trial, sometimes also known as clinical study or research trial, is a process where it tests a potential treatment or a medical device on human volunteers to determine if the treatment or device should be approved for wider use to benefit the general population. Prior to a clinical trial, it is mandatory that the treatment or device must be studied in laboratory animals first to assess the potential toxicity before they can be tried on humans. Only those having acceptable safety profiles and showing the most promise are then allowed to move into clinical trials. It is imperative to note that a "new" treatment or device may not necessarily mean it is "better" for the well-being of individuals such as patients until clinical research and results show otherwise (U.S. Food and Drug Administration 2011). In developed countries like Singapore, clinical trials have become an integral part of a new product discovery and development before it can be brought to the general market (Health Sciences Authority 2011). Unfortunately, unethical behaviour from any person involved in any part of the entire developmental process can have tremendous impact on whether the product will eventually reach the market or not. Known misconduct cases like that of the bioscience researcher, Hwang Woo Suk (Kakuk 2009), has shaken public trust in the research community. This has prompted governments around the world to establish stricter regulations and guidelines for clinical research and to protect participants from unreasonable risks as much as possible. It must be acknowledged that despite all duly diligence, there will be uncertainty inherent in any clinical research. Potential participants in any clinical trial must therefore make independent decision to participate or not only after they have a full understanding of the entire process and the risks that may be involved. Sadly, that may not be true always when facts can be

Table 2.1 It shows the classification of medical devices based on the associated risk level (Health Sciences Authority 2011)

Class	Risk level	Device examples
A	Low risk	Surgical retractors and tongue depressors
B	Low–moderate risk	Hypodermic needles and suction equipment
C	Moderate–high risk	Lung ventilator and bone fixation plate
D	High risk	Heart valve and implantable defibrillator

omitted by staffs involved in the clinical trial. Often, such unethical behaviours are not apparent to the unknowing participants in the clinical trial.

It is important to point out clinical trials involving treatments or drugs and must not be confused as being equivalent to those involving medical devices. First, the definition of medical devices needs to be established. Medical devices are any therapeutic or diagnostic agents or their accessories, which work on non-biochemical mode in the body. From the clinical research perspective, medical devices can be classified into four categories; namely, therapeutic, diagnostic, non-therapeutic and non-diagnostic as well as contraceptive medical devices.

Unlike the four phases of clinical trials involving treatments or drugs, there are typically only three clinical trial phases for medical devices; namely, pilot clinical trial, pivotal clinical trial and post-market surveillance. A pilot clinical trial is an exploratory study limited in size and scope that give insight into the efficacy and safety of a device. However, this cannot provide definitive support for specific mechanistic or therapeutic claims. Defined in this way, pilot trials have been used to guide clinical and translational research for many years (Loscalzo 2009). The intent for a pivotal clinical trial is to establish the safety and efficacy of a device in a statistical representative size of human subjects where the performance of the device is assessed in targeted conditions (Kaplan et al. 2004). As for post-market surveillance, it is the monitoring of the safety and efficacy of a device after it has been released in the market. In the prior two types of clinical trials, the human subjects normally do not have any existing medical condition(s). Thus, it becomes important that post-market surveillance can further confirm the safety and efficacy of a device after it is used in the general population by large numbers of people who can have a wide variety of medical conditions (Resnic and Normand 2012).

To complicate the entire matter, medical devices are also categorised into classes and types based on electrical safety (International Electrotechnical Commission 2012) or risk level (Health Sciences Authority 2011). It is worth noting that such categorisations should be taken into consideration in conjunction with the aforementioned phases of clinical trials. Tables 2.1 and 2.2 illustrate the classifications of medical devices according to the former and latter, respectively. In view of these various classifications, the requirements for clinical trials and the need for one may vary. Further readings on the appropriate clinical trial for a specific medical device and specified application should also be done in context of the guidelines from the local health regulatory bodies.

Table 2.2 It shows the classification of medical devices based on the associated electrical safety (International Electrotechnical Commission 2012)

Type	Details
B (body)	It is the least stringent classification and is used for applied parts that are generally not conductive and can be immediately released from the patient. An example is the probe of a pulse oximeter
BF (body floating)	This is mid stringent classification and is generally for devices that have conductive contact with the patient or have applied parts that are fixed in medium or long-term contact with the patient. An example is the electrocardiogram (ECG) electrode
CF (cardiac floating)	It is the most stringent classification, being required for those applications where the applied part is in direct conductive contact with the heart. An example is cardiac pacing lead

The term applied part is referred to the part of any medical device that comes into physical contact with the patient in order for the device to carry out its intended function

Moral Norms

To further the development of an ethical framework for biomedical research conduct, we can look to the ancient Greek civilisation who expounded the concepts of beneficence and non-maleficence. The provision of benefit and non-harming nature of medical practice has its putative origins in the Hippocratic Oath of 6 BC wherein:

> "...as to disease, make a habit of two things, the provision of benefit and to do no harm".

This means that before any medicinal treatment can be administrated, there is a need for scientific evidences that provides proof of efficacy or "to benefit" and defines the risk-benefit relationship or "do no harm" for specific therapeutic intervention. However, it seems this doctrine has now become a catch-cry of the medical profession, but the underlying concept is a key notion in the evaluation of ethical conduct of human research projects.

One expression of this concept is found in the field of subject selection and randomisation for clinical trials. Randomisation can be defined as being compared to a coin toss that is generated by a computer. The rationale is that no one is certain which treatment (the existing or the proposed) is better and randomisation minimise treatment selection based on the preference of the clinical trial team. Thus, if a proposed therapy was thought or known to have benefit, then the withholding of this treatment to those who may be diagnosed is an unethical act. The approach that should be adopted is that of clinical equipoise first spelled out by Freedman (1987), whereby the population selected for trialling a new device, treatment or diagnostic technique cannot be shown to receive any favour or harm from the intervention outside risks we will discuss later. Additionally, Freedman noted that no interventions outside of the clinical trial can be shown to have a better outcome than those proposed within the clinical trial.

It is worth noting that randomisation may not be always used for clinical trials involving medical devices. In these trials, the accuracy and/or performance of the new or modified device is usually compared against that of an existing device that is used as the reference. There is also another common term used in clinical trials which is known as blinding. The purpose of blinding is to help ensure that bias does not distort the running of the clinical trials or the interpretation of the obtained results. Single-blinding means only the participants do not know whether the standard or new treatment is being administrated on them. For double-blinding, it means that neither the participant nor the conducting staffs know which treatment is administrated. The selection of which treatment to be administrated is predetermined by another staff not conducting that session of the clinical trial. After a pre-specified time, the treatment administrated would then be made known to each participant. Unethical manipulations of the information given to the participants and to the attained data can have adverse effects on the actual outcomes of the clinical trials. Given the possible proximity to the participants, biomedical engineers are in the position to play a significant role to ensure that clinical trials involving medical devices are administrated in an appropriate manner.

Beneficial Versus Non-beneficial Studies

Putting aside the issue of how the actual clinical trial is being conducted, a more fundamental issue then arises, what if the principle of clinical equipoise cannot be applied? Are there not requirements particularly in the field of biomedical engineering where the benefit of a new device or procedure is not yet known? How do we approach the issue of exploratory research where normative values and observations are required before any new development can occur? Clearly, we have cases of research which is "beneficial" that is, research has a clear potential to cure, manage or diagnose an illness, or research which is "non-beneficial" where there is no real or potential benefit to the patient. The treatment of this dichotomy is addressed through our accepted level of risk to the subject. A beneficial study may pose some risk to the subject, which, with their consent may be deemed acceptable. A non-beneficial study would demand lower levels of risk and not compromise the health of the subject outside of the risks of everyday life. How do we argue that most modern-day research is ethical when viewed through such a rigorous framework? Clearly, very few studies would or could be classified as zero risk. Kopelman (2000) points out that risks can be sufficiently low so as to be equivalent to that risk posed by conventional examinations or tests and this may be a risk threshold we can deem to be acceptable.

Authorities have gone further to document guidelines that are in accord with this thinking. The Australian National Health and Medical Research Council et al. (2007) views non-beneficial research as acceptable when the risks do not outweigh those faced in everyday life including that faced in medical examinations. The Department of Health and Human Services of the United States echoes this principle (Department of Health and Human Services 2005).

We are however curiously faced with the dilemma of ever-decreasing risk in conventional practice (that is, risk of harm mediated by tests is falling) which places higher demands on the risk assessments of our research.

The Need to be Involved in the Ethics Process

The question of ethical behaviour may seem as a curious debate to the engineer in this field. Of course one behaves "ethically" as a professional and valued community member. The answer lies in the deeper reflective process of justifying a research project, designing, recruiting and conduct of the study. In practical terms, a common class of study required by the biomedical engineer is the trialling and validation of a new device, therapy or diagnostic technique. A traditional empirical evaluation with mostly qualitative outcomes is seldom adequate in the world of declining health resources. The ethical imperative is fast becoming the design of a study which objectively measures a new interventions effect on minimum number of patients with minimal safety risk. The secondary driver for such behaviour should be the desire to publish such work in peer-reviewed communications, intellectual property restrictions notwithstanding.

The allocation of increasingly scarce healthcare resources has become a predominant role for many biomedical engineers. The ideological framework used to make these judgements will have its origins firmly planted in the field of bioethics. In the assessment of new technologies and having input to policy direction the biomedical engineer must make value judgements that will optimise quality of life while respecting fundamental values. What was once a purely an economic or technical decision in the design process now ventures into the ethics of individual patient care and public health administration.

To assist the decision making process, many professional bodies promulgate a guide or code of ethics to which members subscribe. The Institute of Electrical and Electronics Engineers (IEEE), Engineering in Medicine and Biology Society (EMBS) (2011) provides a code of ethics for biomedical engineers which can be a useful model. Particularly, it addresses the application of the engineering to biomedical applications and policy making. The details of the model are as follow:

Patients and Human Subjects

1. Respect human dignity and privacy of patients and human subjects

Information

1. Ensure proper safeguarding of all confidential information, including information pertinent to patients, subjects, commercial entities, and trade secrets

Environment

1. Promote a culture of cost-effectiveness
2. Support the preservation of a healthy environment

Research

1. Engage in research aimed at advancing the contribution of science and technology to improving healthcare provision
2. Report research results with scientific integrity and proper due credit
3. Observe the rights of human research subjects and strive for a balance between benefits and potential harm
4. Ensure a responsible and human use of animals in research
5. Conduct clinical research studies in accordance with Good Laboratory Practices (GLP) and Good Clinical Practices (GCP)

Profession

1. Hold in high regard the inter-disciplinary nature of healthcare delivery and research. Foster collegial inter-disciplinary relationships. Respect, value and acknowledge the contribution of others
2. Encourage a culture of knowledge exchange and mentorship
3. Avoid or properly disclose conflicts of interest

Research Ethics and Applications

The protection of human subjects in research for biomedical purposes can be seen to have its documented birth in the aftermath of the Second World War. The Nuremburg Trials for Nazi war criminals, specifically the Doctors' Trial Number 23 (Grodin and Annas 1996) saw the revelation of truly abhorrent practices under the name of experiments in concentration camps. Notionally for the advancement of military medicine, the subjects were brutalised and tortured, maimed and murdered. In the formation of a prosecution, the international panel developed a series of ethical principles on which the behaviour of the accused was to be judged. The ten guidelines that then constituted the Nuremburg Code (1947) were:

1. Research participants must voluntarily consent to research participation
2. Research aims should contribute to the good of society
3. Research should be based on sound theory and prior animal testing
4. Research must avoid unnecessary physical and mental suffering
5. No research projects can go forward where serious injury and/or death are potential outcomes
6. The degree of risk taken with research participants cannot exceed anticipated benefits of results
7. Proper environment and protection for participants are necessary
8. Experiments can only be conducted by scientifically qualified persons
9. Human subjects must be allowed to discontinue their participation at any time
10. Scientists must be prepared to terminate the experiment if there is cause to believe that continuation will be harmful or result in injury or death

Historically, the Nuremburg Guidelines were augmented by a major initiative to promote best practice in research ethical decision making and regulation. The Declaration of Helsinki (World Medical Organisation 1996) first muted in 1964 and updated frequently since that time was developed by the World Medical Association. This document sets out the principles of research conduct in the biomedical area in a more comprehensive fashion than the Nuremburg Guidelines were capable of given the limited exposure to technical aspects of medical care.

Some of the principles proposed in the Helsinki Declaration which extend the scope of the Nuremburg Guidelines are:

1. The requirements of independent investigator review
2. Medically qualified personnel to supervise the research and assume the ultimate responsibility for the health and welfare of patients
3. Preservation of results for future review and accuracy
4. Informed consent guidelines
5. Guidelines for research involving children and mentally incompetent persons
6. Evaluation of experimental therapies on patients
7. Importance of determining which medical conditions are appropriate for safe research

It is therefore a commonplace for institutional ethics review committees to require applicants to have read and understood the guidelines promoted by the likes of Nuremburg and Helsinki documents. Contemporaneous guidelines with local interpretations and principles relevant to local practice of customs are also commonly proposed as additional considerations prior to project approval. It is worth reviewing these documents if one is likely to be involved in the preparation of human research ethics applications.

The process of human research ethics application is administratively well documented by local institutions. This may be either the healthcare institution in which the activity will take place or the research institution (university, college or polytechnic) or similar. There is growing trend to the provision of online facilities for submission of a project proposal. Thankfully, acceptance of a common format is also a growing trend in most nations. An example of such a process is seen in the Australian experience (National Health and Medical Research Council, Australian Research Council 2010) whereby the National Ethics Application Form (NEAF) has a standardised structure designed to address all aspects of the scientific, administrative and ethical concerns relevant to a research project. Figure 2.3 shows the typical structure of such a form. All nine sections require a response even if in the negative. As seen, a logical sequence of targeted questions and requests for information is presented.

For each clinical trial that has obtained ethical approval from the local institutional review board or ethics committee, it has been carefully designed to address certain queries pertaining to a new treatment or medical device. Generally, there will be a protocol that maps out what are the procedures, who will be conducting the clinical trial and why there is a need for it. In context of medical devices, they are commonly tested to see how well they are compared to the standard device or

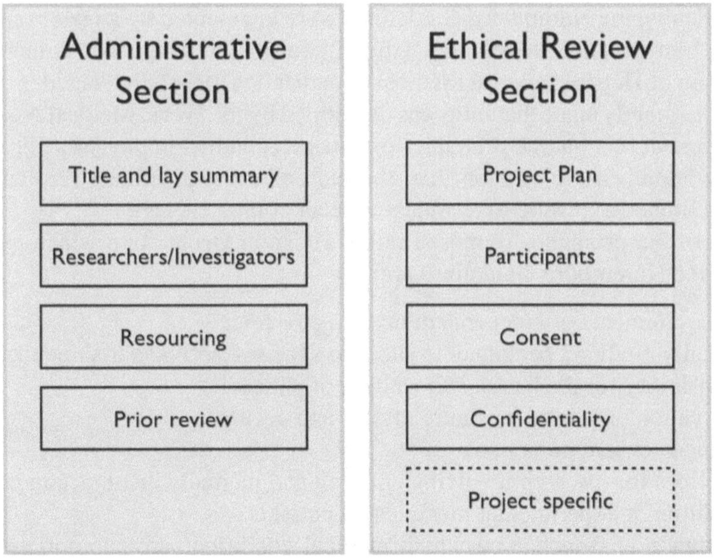

Fig. 2.3 The structure of the National Ethics Application Form (NEAF) application follows standardised approach. Covering all aspects of a project in a comprehensive accessible manner is critical for universal acceptance of this type of application

treatment used. Local regulatory bodies usually provide the general guidelines and assistances to staffs conducting the clinical trial. In addition, parent organisations often offer more extensive technical supports such as helping staffs better design their clinical trials, so that the features of using the new product can be better characterised and reducing the potential risks to those participating in the clinical trial. A typical clinical trial team involving medical devices includes doctors, nurses, biomedical engineers and other healthcare professionals. At the beginning of the clinical trials, the team will check the health status of each participant and assesses the eligibility of that person to participate. Thereafter, those who agree to participate will be given specific instructions and then carefully monitored as well as examined throughout the clinical trial.

Consent and Related Considerations

One key aspect of any applications is the issue of subject consent. As discussed earlier, the historical paternalistic approach was maligned for its lack of respect for the individual. The respect for autonomy of the patient is the primary principle in designing a consent process for most research projects; however, some noted exceptions should be dealt with first.

It is worth noting that some clinical trials will have unpleasant or even serious side effects. While some may be temporary, others can be permanent. Some side

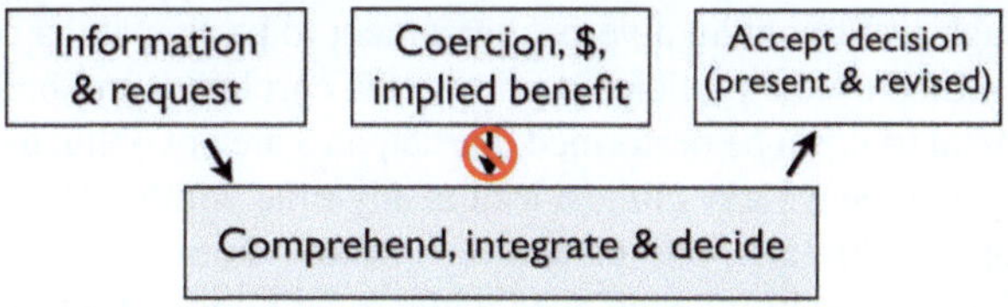

Fig. 2.4 The process of acquiring informed consent is based on information and request followed by a patient decision based on knowledge and experiences. The decision is able to be revised (revoked) at any time without explanation. The consent process should also be seen to be free of direct or implied coercion or inducement

effects can appear during the duration of the clinical trial while others may not manifest even long after the clinical trial is over. Realistically, the degree of risks varies from each clinical trial and the actual health of each participant at the point of the clinical trial. Therefore, it is imperative that all known risks must be fully explained by the team before the clinical trial begins. If new risk information is known during the course of the clinical trial, all participants must be informed. In order to protect the participants, it is a mandatory regulation to obtain informed consent from the participants before they can be involved in a clinical trial. The informed consent process provides an opportunity for the team and participant to clarify the purposes, the risks, the procedures and any doubts pertaining to the clinical trial.

The existence of competence cannot always be expected. Research directed at neonatal or childhood applications, devices or disease states requires the assent of the caregiver and not the patient in question. Similar competency issues arise in older mentally compromised patients or those who may be unconscious either through injury management or trauma. In such rarer cases, a next-of-kin or appointed patient friend has the statutory power to engage in the consent giving process.

The process of acquiring consent has evolved to a three-stage process whereby an invitation for involve and information package is delivered to the target subjects, a period of integration and decision making then occurs followed by the delivery of a decision to the researcher. This is diagrammatically shown in Fig. 2.4 below.

It should be noted that more frequently the standards of communication and periods during which patients can make decisions are increasing. It is not acceptable to present an invitation to involvement with information sheets at one episode of care and expect a decision. Most reasonable committees expect a matter of days to elapse before it is considered reasonable to request an informed decision. This assumes that any and all questions that may have arisen have been addressed by the researchers beforehand.

Obtaining consent from the sick of vulnerable is a circumstance that most ethics reviewers will examine closely. It is clear that a perception of benefit through involvement can arise in this group. It may not necessarily be due to the sense of vulnerability as much as due to a misplaced selflessness, which may override sound reasons not to be involved in a particular study.

It is not enough just to obtain a verbal agreement to participate; a consent form is expected to be signed by the participant prior to the enrolment and before any procedures of the clinical trial can be performed. Participants are not obligated to participant and have the rights to leave any clinical trial at any time without the need to provide for any justification. Similarly, participants also need to know the circumstances that their participation may be terminated even without their consent. In order to ramp up participation rate, unethical omission of certain information about the clinical trial is not uncommon. Sadly, some participants may never get to know about this. Like all other healthcare staffs, biomedical engineers involved in clinical trials are in the position to detect such misbehaviours and flag this out when they witness it.

The immediate benefit of a biomedical engineer being involved in the consent process may not be apparent. However, being involved in best practice activities is an important means of enhancing the reputation of the profession in both the eyes of the public and other related professions. It is of paramount importance that the patient trusts the team of researchers he or she has voluntarily put in a position of power. The allaying of fears and instilling of confidence can help ensure compliance with the requirements of the research project in both the short and long term. This relationship building is critical in conducting long-term follow-up type studies where patient withdrawal can render early work valueless.

Summary

In working with technology that can alter the nature of a living organism or system, the biomedical engineer must consider a framework or set of guiding principles on which to base his or her decision making. Protecting and improving the standard of living is a responsibility for all engineers, not just biomedical engineers. There are historical bases to such codes and progressive development has been forced upon us by the nature of the technology.

A formalised process of research ethics application and approval exists in all areas involved in healthcare research. The structure of such applications is standardised to ensure comprehensive coverage of scientific study design, subject selection and review of hazardous practices and informed consent.

Professional practice has increasingly adopted the moral obligations of medicine as the engineering decision making can have significant ramifications in individual healthcare as well as community-wide health issues.

References

Anson G (1745) A voyage to the South Seas, and to many other parts of the world, from September 1740 to June 1744. Kessinger Publishing, Whitefish, MT

Department of Health and Human Services (2005) Code of Federal Regulations. Title 45 Protection of Human Subjects. US Government Printing Office, Washington, DC

Free C, Phillips G, Felix L, Galli L, Patel V, Edwards P (2010) The effectiveness of M-health technologies for improving health and health services: a systematic review protocol. BMC Res Notes 3:250

Freedman B (1987) Equipoise and the ethics of clinical research. N Engl J Med 317:141–145

Grodin MA, Annas GJ (1996) Legacies of Nuremburg: medical ethics and human rights. JAMA 276:1682–1683

Health Sciences Authority (HSA) (2011) [Internet] [Cited 20 Sep 2011]. http://www.hsa.gov.sg/

Institute of Electrical and Electronic Engineers (IEEE), Engineering in Medicine and Biology Society (EMBS) (2011) IEEE EMBS Code of Ethics [Internet] [Cited 12 Dec 2011]. http://www.embs.org/docs/

International Electrotechnical Commission (IEC) (2012) [Internet] [Cited 11 Jul 2012]. http://www.iec.ch/

Kakuk P (2009) The legacy of the Hwang case: research misconduct in biosciences. Sci Eng Ethics 15:545–562

Kaplan AV, Baim DS, Smith JJ, Feigal DA, Simons M, Jefferys D, Fogarty TJ, Kuntz RE, Leon MB (2004) Medical device development: from prototype to regulatory approval. Circulation 109: 3068–3072

Kopelman LM (2000) Children as research subjects: a dilemma. J Med Philos 25:745–764

Loscalzo J (2009) Pilot trials in clinical research: of what value are they? Circulation 119: 1694–1696

National Health and Medical Research Council, Australian Research Council (2010) National Ethics Application Form (NEAF) version 2.0 [Internet] [Cited 12 Dec 2011]. www.neaf.gov.au

National Health and Medical Research Council, Australian Research Council, Australian Vice-Chancellors' Committee (2007) National statement on ethical conduct of human research [Internet] [Cited 10 Dec 2011]. http://www.nhmrc.gov.au

Nuremburg Code (1947) In trials of war criminals before the Nuremburg military tribunals under control council law, Number 10, vol 2. US Government Printing Office 1949, Washington, DC, pp 181–182

Resnic FS, Normand SL (2012) Postmarketing surveillance of medical devices – filling in the gaps. N Engl J Med 366:875–877

U.S. Food and Drug Administration (2011) Inside clinical trials: testing medical products in people [Internet] [Cited 12 Dec 2011]. http://www.fda.gov/Drugs/ResourcesForYou/Consumers/ucm143531.html

World Medical Organisation (1996) Declaration of Helsinki. Br Med J 313:1448–1449

Chapter 3
Ethical Considerations in Clinical Engineering

Winston Gwee

Keywords Clinical engineer • Public health • Patient safety • Product design • Preventive maintenance • Equipment management • Conflicts of interests • Whistle blowing

Introduction

To many, clinical engineering is often confused as being the same as biomedical engineering. In particular, the former is actually a branch of biomedical engineering that manages the deployment of medical technology and integrates it appropriately with desired medical practices. Typically, a clinical engineer works in a healthcare establishment such as a hospital or a specialists' center. The clinical engineer can be considered as a professional who would bridge the communication gaps amongst the medical, administrative, and technical personnel in the healthcare sector. From this regard, the work undertaken by a clinical engineer would have a direct impact in improving the care for patients by leveraging technological solutions in the diagnosis and therapy. By the definition of the American College of Clinical Engineering (ACCE), a clinical engineer is a professional who supports and advances patient care by applying engineering and management skills to healthcare technology (American College of Clinical Engineering (ACCE) 2011). Figure 3.1 shows that there are a number of career options (though not an exhaustive list) for a biomedical engineer which includes being a clinical engineer.

It is known that the healthcare delivery system is a very complex environment where facilities, equipment, materials, and a full range of human interventions (such as patients and staff) are involved. This may lead to unacceptable risk when

W. Gwee (✉)
Electronic and Computer Engineering Division, School of Engineering,
Ngee Ann Polytechnic, 535 Clementi Road, Singapore, Singapore 599489
e-mail: gcs@np.edu.sg

J.Y.A. Foo et al., *Ethics for Biomedical Engineers*,
DOI 10.1007/978-1-4614-6913-1_3, © Springer Science+Business Media New York 2013

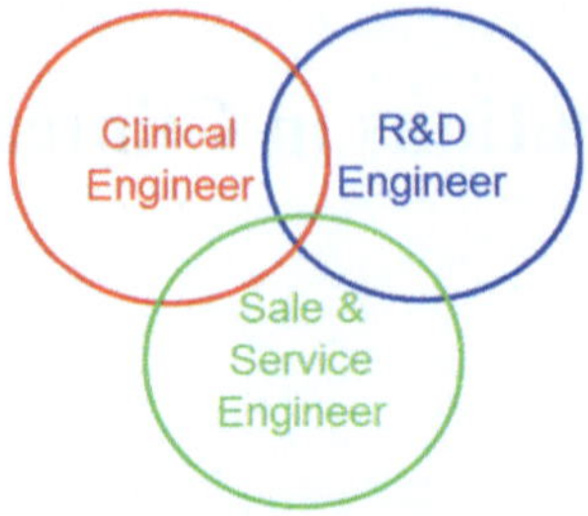

Fig. 3.1 Clinical engineer is sometimes regarded as being a biomedical engineer in some organizations. However, it is just one of the many career options for a biomedical engineering graduate. It is worth noting that the above figure does not show the exhaustive list of these career options

programs for monitoring, controlling, improving, and educating all involved are not appropriately integrated by qualified professionals (Dyro 2004). Therefore, an understanding and awareness of the various factors that may affect public health and patient safety are of importance to the clinical engineer's work. Besides being competent, clinical engineers must also be ethical in their approach in performing their daily work. This is understandable since the work performed by any clinical engineer would have a direct impact on the well-being of patients. In this chapter, how the role of a clinical engineer, safety threats in a healthcare establishment, and the value of ethics can intertwine in this interdisciplinary profession will be discussed.

Job Scope of a Clinical Engineer

It is important to first understand the job scope of a clinical engineer (which differs from that of a biomedical engineer in some aspects) before the unique interplay of ethics and the clinical engineering profession can be better appreciated. A clinical engineer is responsible for installing and maintaining medical equipment in hospitals or healthcare facilities. Clinical engineers sometimes have to train doctors, nurses, and allied medical personnel who will be using the equipment. It is noted that, with changes to the clinical engineering industry, an increasing number of clinical engineers work for contract service agencies and as independent contractors (Dyro 2004). Figure 3.2 illustrates the multifaceted role played by clinical engineers. Simply, they must interface with the clinical staff, vendors, hospital administrators, regulatory agencies, and even the patient (in some cases) to ensure that the medical equipment within the hospital is safely and effectively utilized.

Specifically, the major functions of a clinical engineer are to:

- Perform the following on clinical equipment owned and/or used within the health system in compliance with regulatory agencies:

  (a) Inspection of all incoming equipment (i.e., both new and returning repairs)
  (b) Installation
  (c) Preventive and corrective maintenance
  (d) Special request service

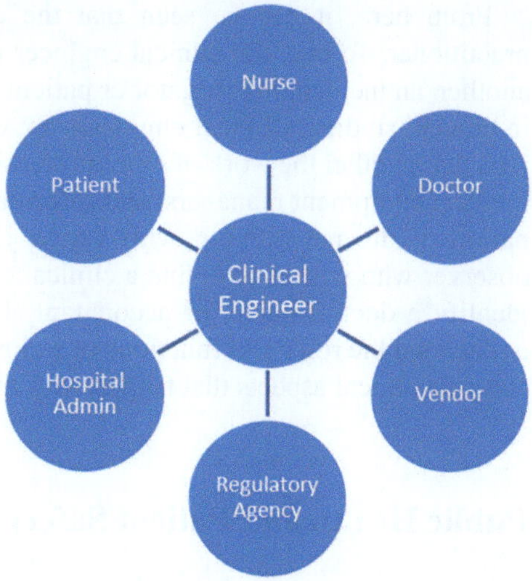

Fig. 3.2 People whom the clinical engineer interacts within hospitals or healthcare facilities

- Provide prepurchase evaluations of new technology and equipment
- Assist clinical departments with service contract analysis, negotiations, and management
- Provide cost-effective management of a medical equipment calibration and repair service
- Coordinate clinical equipment installations including, planning, scheduling, and oversight
- Research equipment issues for health system professional and administrative staff
- Conduct device incident investigations
- Coordinate outside engineering and technical services performed by vendors
- Train medical personnel in the safe and effective use of medical devices and systems
- Perform clinical applications engineering, such as custom modification of medical devices for clinical research, evaluation of new noninvasive monitoring systems, etc.
- Develop and implement documentation protocols required by external accreditation and licensing agencies

In addition, the knowledge, skills, and abilities of a clinical engineer include:

- Knowledge of human anatomy, physiology, electronics, and electro-mechanical fundamentals, medical equipment operation and troubleshooting, and safety in healthcare facilities
- Analytical skills to determine the cause of equipment malfunction or failure
- Practical skills to do repair and perform preventive maintenance of electro-mechanical equipment
- Ability to write reports and make clear presentations on technical and operational issues
- Interpersonal skills to work effectively with clinical staff, vendors, and fellow engineers

From here, it can be seen that the clinical engineer is a multidisciplinary practitioner. To one, the clinical engineer is an advisor on technology selection; to another, an incident investigator or patient-safety expert; to a third person, a partner in clinical studies. Clinical engineers are everywhere on the health scene, often in jobs that parallel the work of other professionals, including medical physicists, biomedical equipment managers, and information technology specialists. This combination of diverse training with varied job experiences can confuse the casual observer who wishes to define a clinical engineer in the same way that one would identify a doctor, nurse, or accountant (Dyro 2004). Thus, it is essential to first understand the roles and functions of a clinical engineer before one can fully appreciate the ethical aspects that tagged with this profession.

Public Health and Patient Safety in Healthcare Facilities

Clinical engineers, doctors, nurses, and other allied medical personnel must continue to work to strive for patient safety. These personnel must be aware of the main sources of safety threat to patients, public (visiting family members of patients or contractors), and staff. Thus, knowing these threats can better help a clinical engineer understand the implications of compromising the safety measures and thereby, minimizing others from being exposed to these threats. Being unethical would mean placing others in a position that is detrimental but benefitting one's own self, whether tangibly (such as monetarily) or intangibly (such as saving time or lesser work). The ten main sources of safety threats are:

1. Earth, air, fire and water
2. Chemicals and drugs
3. Microorganisms and vermin
4. Waste
5. Sound and radiation
6. Electricity
7. Natural and unnatural disasters
8. Surroundings
9. Gravity and mechanical stress
10. People

Earth, Air, Fire, and Water

Earth (dirt and dust) can contaminate wound sites and damage delicate electronic instrumentation and computer data storage devices. It can also harbor pathogenic organisms.

Air pollution can adversely affect air supply that is needed to provide respiratory support and to power pneumatic devices.

Fire is particularly hazardous when a person's ability to evacuate is compromised by illness or disability.

Water can cause damage and compromise patient safety. For example, sterilization barriers are breached when high-humidity environments cause water accumulation on sterile wrappers. Water can also damage electrical equipment. An example is that excessive accumulation of water in respiratory therapy devices can impair efficacy of these devices.

Chemicals and Drugs

Many potentially injurious chemical compounds have been associated with the operation of medical devices and instrumentation. An example is the thermal decomposition of ether to formaldehyde in infant incubators by the catalytic action of high-surface-temperature metal heating element. Other examples are ethylene oxide, glutaraldehyde, and peracetic acid used for sterilization.

Drugs and medication errors constitute one of the major sources of hospital incident reports. This can easily be caused by unintended oversight but the implications and complications can be substantial (Channel New Asia 2009).

Safety problems include:

- Adverse drug reactions
- Inappropriate dosage
- Inappropriate drug
- Inappropriate frequency of administration
- Inappropriate route of administration

The incompatibility of certain drugs with the plastics of infusion pumps is a source of risk. It has been shown that nitroglycerin in solution administered by infusion pumps interacts with certain formulations of polyvinyl chloride tubing, causing the problem of inaccurate dosage.

Microorganisms and Vermin

The three major complications of invasive monitoring are:

- *Infection*. Growth of a parasitic organism within the body
- *Thrombosis*. Formation or presence of a blood clot in a blood vessel
- *Embolism*. Sudden interruption of blood flow to an organ or body part due to a clot

Infection can be spread through:

- Medical devices
- Vermin (examples: lice, fleas, and rodents)
- Animals
- Fomites (examples: bedding and clothes)
- Food
- People

Waste

Waste can be in the form of solid, liquid, or gas. Its disposal can be posed as risks for hospital staff and patients as well as the community in which the hospital is situated. Nurses have been stuck by needles in the course of injecting a patient or disposing of a used needle. Housekeeping personnel have been victims of improperly discarded needles, broken glass, biohazardous, and radioactive waste. Chronic low-level exposure to inhalation anesthetics is associated with spontaneous miscarriages, liver disease, cancer, and other physiological disorders. Waste anesthetic gas exposure can be hazardous.

Generally, hazardous waste that is no longer used can be either recycled or thrown away. Due to the disposal costs, this does not happen until sufficient quantities of hazardous waste are collected. Moreover, there are concerns if bad practices such as pouring the hazardous waste down a common drain or sink do exist. By right, any substance to be discarded through the conventional drainage system, approval must first be obtained from the local public-own treatment works stating that they will accept the material in question as influent, and that the quantities to be discarded are within the accepted range for the time frame involved. However, violations to this good practice may not be easily detected.

Sound and Radiation

High levels of sound can injure the auditory system. Infants are more susceptible to the effects of sound; hence, attention should be paid to noise levels within infant incubators. Many devices utilize ultrasound. Under certain conditions, the intensity of the ultrasonic beam in some pulsed or continuous-wave ultrasonic devices may approach or exceed guideline levels for known biological effects.

Ionizing radiation hazards are well-known. A good radiation safety program is essential to ensure that radiographic equipment and protective measures in radiographic suites meet acceptable performance and safety standards. A large percentage of accidents or incidents related to radiographic equipment are mechanical in nature and not related to radiation injury.

Nonionizing radiation may have significant health hazard in hospitals. It includes ultraviolet, microwave, and laser radiation. The eye is the most susceptible organ from this type of radiation. Ultraviolet therapy is employed in the treatment of some skin disorders as well as treating jaundice. Ultraviolet rays can cause erythema (reddening) of skin, keratitis, and skin carcinogenesis. Microwave radiation is commonly used in physical therapy for diathermy treatment of patients. Microwave effects are largely thermal, with the eye being the most susceptible (where it can cause cataractogenesis). Microwave devices that are leaky should be periodically checked to ensure that levels of radiation are safe.

Lasers are popularly employed as surgical tools. It is particularly hazardous as the beam can travel long distances with little attenuation and can reflect off surfaces in a room. The intensity of the beam is sufficient to burn body tissues as it is required in surgical procedures. Momentary eye contact with the beam can cause severe eye damage. A laser safety program is recommended whenever lasers are used. Eye protection and restriction of the area to trained personnel are essential.

Electricity

Hospitals have expended considerable resources on ensuring an electrically safe environment. The estimated risk of a fatal micro-shock is about the same as being struck by a meteorite. However, the performance of medical devices can be adversely affected by irregularities in electrical power distribution systems. Line voltage variations, line transients, and power interruption can result in harm to the patient by affecting the performance of such devices as apnea monitors, ventilators, electrosurgical units, and electrocardiographs. Electromagnetic interference is another source of risk (especially with the use of mobile phones). Clinical engineers must diligently perform all the required electrical safety tests of medical equipment as detailed in the preventive maintenance program. This has to be done properly without skipping any test either due to heavy workload or simply indifferent attitude.

Natural and Unnatural Disasters

Natural disasters can constitute tornado, hurricane, and earthquake while unnatural disasters can be a nuclear power plant meltdown. Hospital evacuation planning in catastrophic and emergency situations will minimize loss in these circumstances. An example would be the outbreak of severe acute respiratory syndrome (SARS) in Singapore from 1 March to 11 May 2003 where various national prevention and control measures were undertaken to control and eliminate the transmission of the infection (Deurenberg-Yap et al. 2005). During such incidences, it would seriously overtax or threaten to overtax the routine capabilities of a healthcare establishment. This situation creates the need for emergency expansion of facilities and operation of the facility in an unfamiliar environment and pace. Thus, there should be emergency plans not just for a single disaster but separate plans for different disasters. It is becoming increasingly possible for multiple disasters (both natural and unnatural) to occur concurrently such as the earthquake and tsunami off the Pacific coast of Tohoku, Japan which later triggered the nuclear plant meltdown at Fukushima, Japan.

Surroundings

Patients and staff can suffer stress from their surroundings. Dehumanization of patients can, following a heart attack, lead to depression, denial, and dependency, collectively known as "coronary madness."

There are seen or unseen dangers associated with the presence of medical devices. An example of the unseen danger is the fringe effects of the high-strength magnetic field of the magnetic resonance imaging (MRI) machines.

Gravity and Mechanical Stress

There are a number of hospital injuries reported that are due to slips and falls which also include infants having been injured in falls from incubators. Furthermore, mechanical stress can also injure patients and staff. It has been reported that monitors have fallen from overhead shelves onto staff (ECRI Institute 1992). Other examples can include pediatric patients being trapped in crib side rails due to poor design, and excessive pressure against parts of the body during long surgical procedures has been implicated as a cause for pressure sores. Patients undergoing anesthesia may also have broken teeth and vocal cord damage during intubation. Like many other professions, clinical engineers can have common back injuries when they are not careful when lifting heavy medical equipment or adopting a position that is strenuous to the body.

People

This is the most significant factor in terms of number of incident reports and malpractice claims. Hospital personnel who do not follow proper patient care procedures or surgical procedures, who fail to monitor, or who do not know how a medical device is designed to work pose safety problems for the patient. Poor communication between doctors and nurses can adversely affect a patient's chance of recovery. Understanding of and respect for the role of each member of the healthcare team promote between cooperation and an overall safer environment for the patient.

Physicians and nurses are obligated to use technology that has been accepted as the standard of care, and liability is incurred if available technology is not used or if available information is not acted upon. Failure to diagnose correctly can be related to the presence, quality, and limitations of diagnostic data obtained via the medical device. Other patients, staff, visitors, and salesmen can cause injury to patients. Salesmen or visitors who have observed surgical procedures have been sued to compensate the patient for psychological harm caused by the invasion of privacy.

Conflicts of Interests

The clinical engineer may be subjected to conflicts of interests during the procurement, rental, or service contract of medical equipment for the hospital. Here are some examples:

- Purchases or rental from vendor who is a friend or relative with the intention of receiving kickbacks (e.g., monetary or material rewards, favor, etc.).
- Purchases or rental from the clinical engineer's close relatives[1] or from a business owned by his close relatives or himself.
- Revealing the bids of other competitors to benefit the intended vendor.
- Being treated to a meal or accepting gift from the vendors. Acceptance of anything can create the appearance of conflict of interest.

Ethical Issues in Design and Manufacture

Ethical questions may arise in the different phases of a product design as shown in Table 3.1.

Ethical Issues in Equipment Management

Equipment Acquisition Phase

This is the process by which the hospital introduces new technology into its operation. The process should involve every clinical and support department.

Clinical engineers who are involved in procurement activities have a responsibility to behave ethically at all times. Ethical behavior supports openness and accountability in a procurement process. Ethical behavior can also reduce the cost of managing risks associated with theft, fraud, corruption, and other improper behavior and enhance confidence in hospital administration.

[1]Close relative usually refers to spouse, parent, child, sibling, grandparent, grandchild, or in-laws and step-relatives in the same relationship.

Table 3.1 Typical ethical questions associated with product design

Steps in product design	Possible ethical questions
Market study	Is the study unbiased? Is the study comprehensive? Has it been exaggerated to attract investors or management support?
Conceptual design	Will the product be useful or will it be just a gimmick?
Embodiment design	Does the design team ensure that the computer programs are tested comprehensively and work reliably? Are there any patents or copyrights being violated?
Detail design	Are the test results checked thoroughly? Are the materials selected toxic?
Manufacturing	Is the workplace safe and free of environmental hazards? Is child labor exploited or slaves employed?
Product use	Is the product safe to use? Will there be serious errors as a result of misuse or mishandling of device? Are users informed of possible hazards?
Retirement from service	Are the materials and associated disposables selected to allow for recycling or reuse?

Outline of Processes

1. Justification for the new equipment
   - Assessment of the need
   - Proposal should include:
     (a) Clinical considerations
     (b) Financial considerations
     (c) Environmental considerations
   - Budget needed
2. Selection of the new equipment usually includes the following:
   - Literature review of the new equipment can include:
     (a) Library
     (b) Subscriptions
     (c) Standards
     (d) Manufacturer's literature
   - Proposal from interested vendors
     (a) Details of submitted proposal
     (b) Preliminary review to ensure minimum requirements are met
   - Engineering testing of the equipment
     (a) Compliance with mandatory safety test
     (b) Compliance with expected performance test

- Clinical testing of the equipment
  (a) Trial use in the expected application
  (b) Questionnaire or interview end users
  (c) Contact other possible users
- Assessment
  (a) Ranking of equipment based on engineering and clinical requirements
  (b) Requests for quotations
  (c) Final selection to be made
- Negotiate
  (a) Ensure all details are finalized before the contract is signed

3. Implement the purchase of new equipment
   - Generate the purchase order to effect the process
   - Agree on the delivery date and installation matters
   - Perform acceptance testing before taking over the equipment
   - Vendor to provide the following training:
     (a) Operator
     (b) Service
4. Conclude the procurement process
   - Generate a report that encompasses the entire process
   - Follow-up on any outstanding matters (if any)

Equipment Control Program

This involves the management of medical devices and equipment within the healthcare facility. It begins with the receipt of newly acquired equipment until its end of life cycle. Good ethical values are needed in the processes throughout the equipment control program. An example where ethics may be compromised is during the negotiation process with vendors for service contracts. In most large healthcare establishments, a clinical engineer is usually assigned to facilitate the equipment control program and may be known as the equipment control manager.

A carefully planned equipment control program is required to ensure that:

- Equipment that is appropriate and adequate for the need is purchased
- Equipment is properly maintained
- Equipment is included in a planned replacement schedule
- Equipment is monitored for liability risks
- Staff is adequately trained in equipment use

Responsibilities of Equipment Control Manager

1. Review all equipment purchase requests
2. Coordinate prepurchase evaluation of clinical equipment
3. Coordinate incoming inspection testing of new clinical equipment and equipment repair
4. Ensure that new clinical equipment is tagged with an equipment control number and properly entered in the inventory
5. Ensure that operator and service manuals are obtained for each clinical device
6. Establish periodic calibration and performance and safety testing of clinical equipment, electrical receptacles, electrical beds, and isolated power distribution systems
7. Establish an equipment repair program
8. Review requests for vendor service contracts agreements
9. Monitor vendor service and contracts
10. Coordinate training of technical personnel in the proper and safe use of equipment
11. Coordinate training of clinical personnel in the proper and safe use of equipment
12. Provide a means of advising clinical personnel of the status of their equipment
13. Ensure that the hospital is in compliance with local, state, federal, and other regulatory agency requirements concerning the control of clinical equipment
14. Review equipment and product hazard reports, alerts, and recalls and ensure that appropriate follow-up action is taken
15. Investigate all equipment-related incidents

Besides the equipment control manager, other clinical engineers must also ensure that the following aspects of equipment control program are carried out with due diligence and ethically.

Inventory

Why is there a need for an inventory control system within a healthcare establishment? The reasons include:

- To have a record of the quantity and location of specific items
- To ensure that all the equipments are included in the preventive maintenance schedule
- To provide a means to document repairs and maintenance for each item and their costs
- To provide a resource for determining new and replacement equipment needs
- Required even if all the equipment in the hospital is serviced by outside sources

Each piece of equipment should have an equipment control record with the following information:

(a) Equipment (item) identification number
(b) Item name
(c) Model number
(d) Serial number
(e) Equipment manufacturer
(f) Location of item
(g) Owner department
(h) Acquisition date
(i) Acquisition cost
(j) Warranty expiration date
(k) Vendor
(l) Identity of who is responsible for maintenance (in-house, outside vendor, shared-service, etc.)
(m) Inspection schedule

Hazards

The equipment control program should have an equipment hazard awareness program to minimize risks to patients and reduce the risk of liability to the institution. This program should be substantiated by written policies and procedures. A special individual should be designated as the coordinator for processing hazards alerts and notifications, and there must be an effective means for documenting that appropriate action was taken.

Quality Assurance

Quality assurance in equipment control can be defined as planned and systematic activities implemented in a quality system so that its requirements for equipment will be fulfilled. Similarly, a clinical engineer needs to perform the responsibility in an ethical manner. Thus, quality assurance in equipment control can involve:

(a) Proper selection of equipment:

- Evaluate devices from several manufacturers
- Engineering evaluation
- Clinical evaluation
- Check performance, safety, ease of use, training, and service

(b) Acceptance testing where an incoming inspection should include:

- Inspection for physical damage
- Verification of all items and supplies (including operators and service manuals)
- Assembly of equipment
- Electrical and mechanical safety testing and inspection
- Performance assurance testing
- Documentation of inspection
- Processing of inventory control information including tagging of equipment control number and scheduling of preventive maintenance
- Arranging for in-service training of the clinical engineering (as needed) and clinical staff prior to use

(c) Maintenance

- Safety and performance tests at intervals specified by manufacturer, but at least every 6 months.
- Test, calibration, and inspection protocols should be written or available for each instrument.
- Appropriate documentation of scheduled inspections.
- Program should be reviewed periodically to ensure that the inspections have been carried out, the schedule is appropriate, the test procedures, calibration of test equipment, and competency of testing personnel are adequate.
- Verification of service by outside organizations must also be documented. It is best not to assume that others will do the proper documentation.

(d) Repairs

- Initiated and completed in a reasonable time by competent, adequately trained personnel
- Vendor service must be documented and monitored by the clinical engineering staff
- Periodic reports should be sent to user departments on the status of their equipment

(e) User training

- Hospitals continually procure new and replacement equipment; they rent or borrow equipment for temporary use.
- Changes in personnel are also common.
- There is a need for continuing in-service training of clinical staff on safe and proper use of equipment to remind staff.
- Should have written procedures to ensure that the content and frequency of sessions are adequate; all shifts are covered; participation by individuals is documented; suitable informational resources are available; adequate follow-up opportunities for all staff exist.

(f) Management

Managing an equipment control program includes:

- Implementing all the essential elements of the program
- Monitoring the program for quality and appropriateness
- Making adjustments to ensure that the changing needs of the institution are satisfied
- The goal should be to ensure that utilization of technology is optimized. The program should be evaluated for cost, user satisfaction, scope, effectiveness, and efficiency. The efficiency of the program is a measure of its cost or effectiveness as compared to alternative service options.

(g) Cost

The principal goals of an equipment control program are to effect quality assurance and enhance the risk management position of the hospital in matter related to patient care equipment. An objective of the program should be to achieve these goals by the most economical means without compromising the mission. Equipment selection and maintenance have direct impact on cost.

Equipment Maintenance

The goal of a biomedical equipment maintenance program is to provide safe, calibrated, and operational equipment for delivery of the best health care possible. An effective program should reduce the inconvenience and frustration caused by malfunctioning equipment and the time lost because of unavailability of equipment. Good ethics must again be observed here so that all equipments are well maintained and that shortcuts or malpractices are not allowed during preventive maintenance.

Preventive Maintenance

1. Problems affecting an effective preventive maintenance or PM:
   - *Equipment unavailable.* Especially when it is used for daily monitoring or in critical care areas like intensive care unit
   - *Equipment cannot be located.* It can be in transit between departments
2. Inspection schedules:
   - Desired testing interval should not exceed 6 months.
   - This period can be extended if proper evaluation is done and proper approval is obtained.
   - Initial schedule can be established using manufacturer's recommendation and the experience of others.

- If preventive maintenance is performed too often, valuable time and money are wasted. However, if inspection interval is too long, then maintenance is not effective.
- The preventive maintenance schedule should be reviewed periodically to determine if the schedule is meeting the requirements of the equipment.

3. Two levels of preventive maintenance:

   - For major level, it is usually extensive and not frequent (i.e. annual or semiannual).
   - For minor level, it is usually performed quarterly, monthly, or even weekly.

4. Four common procedures of preventive maintenance:

   - *Visual inspection*. Identify any loose hardware, dented or damaged cases, flaking paint, illegible labels, loose components, fluid leaks, dirty fans, etc.
   - *Cleaning*. Standard procedures that state the solvents or cleaning agents that can be used as well as the cleaning methods.
   - *Function testing*: This is probably the most important step that will guarantee the operational effectiveness and calibration of the equipment.
   - *Safety testing*: Perform electrical safety in terms of level of leakage currents and quality of grounding; pressure release valves; protective coverings or shields; emergency mechanical releases; and emergency electrical or gas shutoffs.

Negligence

Clinical equipment users sometimes mandate the clinical engineers to repair or calibrate the equipment within a short notice. This causes stress on the clinical engineers which may lead to negligence or taking shortcuts. An example is to assume that the equipment would pass the preventive maintenance checks by simply verifying that it functions after a repair. Equipment that appears to function well may, for example, possess excessive leakage currents on the patient leads or applied parts.

Another situation that may arise is the shortcut taken during the performance verification checks of say 40 new ventilators by one engineer to be completed within a short timeframe. He might irresponsibly assume that the rest would work after tirelessly ensuring that the first few ventilators have passed the performance verification checks perfectly.

One overlooked device is the hospital bed where shoulder bolts supporting the four corners can become very loose. These bolts can fall out, making the beds uneven and unstable and presenting a hazard to both patients and staff. Clinical engineers performing preventive maintenance may focus on the electrical safety aspects and overlook the mechanical parts of the bed.

The ECRI provides a complete list of the hazard reports, user experience network reports, guidance articles, posters and checklists, and frequently asked questions (FAQs) that make up medical device safety reports. All these details can be obtained from the following website (http://www.mdsr.ecri.org/index/index.aspx).

Major Challenges of Clinical Engineers

It has been reported that some of the major challenges facing today's clinical engineers (Loughlin and Williams 2011) include:

1. More and more medical devices are networked, thus requiring increasingly complex management procedures that are integrated with IT and risk management protocols. For example, a patient monitor that works fine when it is standalone and not plugged into the network, but might malfunction once it is plugged in. A clinical engineer who is an expert on the device, but does not have networking troubleshooting skills is at a loss and unable to resolve the problem.
2. The rise of wireless devices brings with it a new set of design, education, training, and maintenance issues, as would be the case for any other class of device. This could lead to the need for drastic changes in infrastructure to meet the bandwidth demands.
3. With increased medical equipment inventory comes increased preventive maintenance workload. The problem is mainly due to the different vendors and different models of the same equipment. Does the engineering department increase number of staff and/or increase training hours for these engineers?
4. Increasing use of monitors by patients at home for self-monitoring may result in a bigger issue for preventive maintenance due to mishandling by patients or family members and children.

Code of Ethics

It can be defined as a statement encompassing the set of rules based on values and the standards of conduct to which practitioners of a profession are expected to conform (The Free Dictionary 2012).

Code of Ethics of Engineers

The Fundamental Principles

Engineers uphold and advance the integrity, honor, and dignity of the engineering profession by:

1. Using their knowledge and skill for the enhancement of human welfare
2. Being honest and impartial, and serving with fidelity their clients (including their employers) and the public
3. Striving to increase the competence and prestige of the engineering profession

The Fundamental Canons

1. Engineers shall hold paramount the safety, health, and welfare of the public in the performance of their professional duties.
2. Engineers shall perform services only in the areas of their competence; they shall build their professional reputation on the merit of their services and shall not compete unfairly with others.
3. Engineers shall continue their professional development throughout their careers and shall provide opportunities for the professional and ethical development of those engineers under their supervision.
4. Engineers shall act in professional matters for each employer or clients as faithful agents or trustees and shall avoid conflicts of interest or the appearance of conflicts of interest.
5. Engineers shall respect the proprietary information and intellectual property rights of others, including charitable organizations and professional societies in the engineering field.
6. Engineers shall associate only with reputable persons or organizations.
7. Engineers shall issue public statements only in an objective and truthful manner and shall avoid any conduct that brings discredit upon the profession.
8. Engineers shall consider environmental impact and sustainable development in the performance of their professional duties.
9. Engineers shall not seek ethical sanction against another engineer unless there is good reason to do so under the relevant codes, policies, and procedures governing that engineer's ethical conduct.
10. Engineers who are members of a society shall endeavor to abide by its constitution, by-laws, and policies of the society, and they shall disclose knowledge of any matter involving another member's alleged violation of the code of ethics or the society's conflicts of interest policy in a prompt, complete, and truthful manner to the chair of its committee and ethical standards and review.

Procedure for Solving Ethical Conflicts

1. Internal appeal option
   (a) Individual preparation
      - Maintain a record of the event and details
      - Examine the company's internal appeals process
      - Be familiar with the state and federal laws that could protect you
      - Identify alternative courses of action
      - Decide on the outcome that you want the appeal to accomplish

   (b) Communicate with your immediate supervisor
      - Initiate informal discussion
      - Make a formal written appeal
      - Indicate that you intend to begin the company's internal process of appeal

(c) Initiate appeal through the internal chain of command
   - Maintain formal contacts as to where the appeal stands
   - Formally inform the company that you intend to pursue an external solution

2. External appeal option
   (a) Individual actions
      - Engage legal counsel
      - Contact your professional society

   (b) Contact with your client (if applicable)
   (c) Contact the media

Closing Thoughts

The need for clinical engineers has been increasing due to the growth and diversity of technological applications in the healthcare sector. This has resulted in clinical engineers having a more significant role within a healthcare establishment and the need for them to be more conscious of their responsibilities to the public as well as the patients. It is no longer just a matter of professionalism but clinical engineers may be caught in dilemmas with ethical issues to manage. It is then critical for the clinical engineers to understand their role well and that their contributions are recognized by other healthcare professionals. More importantly, clinical engineers themselves need to know the seriousness of their responsibilities as they have a multifaceted role (medical, administrative, and technical) to contribute to the healthcare system. Particularly, they contribute to medical equipment maintenance, use and develop instrumentation, fulfill administrative responsibilities, resource planning and management, training and education as well as research. In short, clinical engineers (being part of the healthcare team) must be prepared to face ethical issues arising from public health, patient safety, conflict of interest, equipment management (defective or inadequate equipment), confidentiality, and palliative care.

Discussion Questions

Discussion Question 1

Explain why it is important for a clinical engineer to be ethical.

Discussion Question 2

For each of the following phases, identify and discuss which of the processes the clinical engineer's ethics could be compromised. Give examples.

(a) Equipment acquisition
(b) Equipment control program
(c) Equipment maintenance

Case Studies

Case Study 1

Company A's CEO decides to offer a cash reward to whistle blowers in an attempt to improve its corporate governance and eradicate any fraudulent activities in the company. It will reward up to US$38,000 cash to any employee, supplier, business associate, or member of the public who informs the company of any wrongdoing by any of its executive directors or senior managers, if the information leads to an admission of guilt or successful prosecution. If the wrong-doing is committed by a lower level employee, the reward is smaller. The CEO said he hopes to root out any wrongdoing, be it cheating, stealing, or taking bribes. This company also made headlines when it barred its top executives and key finance personnel from the local casinos.

Discuss the pros and cons of this approach to whistle blowing.

Case Study 2

[This case was adapted from (Darr 2005)]

Mary is the head of the radiology department of XYZ Hospital. She is going on 6-month maternity leave and has appointed John as the acting head. He would be in charge of two technicians, an administrative executive and $400,000 equipment. He would have the authority to purchase radiographic supplies that are obtainable from three companies which the hospital had bought for years.

As Mary oriented John, she told him how much she liked the meetings with sales representatives of the three vendors. Over the years, one of them had become a close personal friend. She told John that most meetings were held at nice restaurants. If the meetings were held at her office, the sales representatives would bring along "a little something" like perfume, bottle of brandy, a pen set, etc. She estimated the cost of the gift to be between $40 and $50.

When John asked Mary whether there was a policy about accepting gifts from vendors, Mary was upset by the question. She responded curtly that the hospital trusted its managers and allowed discretion in such matters. John then asked if accepting gratuities might suggest to other staff that her decisions were influenced by the pecuniary relationship with the sales reps. Mary was angry: "I know you think what I'm doing doesn't look right. That's not fair! I work long hours as a manager. It takes effort and time to order and maintain proper inventory. If things go wrong, it's my neck in a noose. My work has been exemplary."

Discuss what John should do when faced with similar gifts from the sales representatives.

Case Study 3

Aston and Peter have been working as clinical engineers in a local hospital for 3 years. Both of them have become very close and consider each other as buddy. One day, three electrocardiogram (ECG) monitors are due for routine preventive maintenance (PM) and calibration. However, all three monitors are released to the clinical engineering department rather late in that day due to a training course. Both Aston and Peter are very familiar with the operation and performance of these ECG monitors. As Aston has an important personal appointment in the evening, Peter has agreed to service two of the ECG monitors while Aston services one. After a quick visual check, Aston tells Peter that the ECG monitor he is servicing looks fine and is going to update the PM status as "completed" before printing the PM sticker to be placed on the ECG monitor.

(a) How should Peter handle the situation?
(b) Identify and describe the ethical issues involved.

Case Study 4

A call for tender is a key process taken by hospitals when there is a need to purchase medical equipment. The process includes inviting interested vendors to submit their recommended equipment with all details such as costs, documentations, description of acceptance testing, etc. Mark works for a hospital that is going to purchase a new series of magnetic resonance imaging (MRI) machines and he is assigned as the key staff to handle the process. During a weekend gathering, a personal friend of his who also happens to be a medical equipment vendor causally asks Mark about the tender.

(a) How should Mark handle the situation?
(b) Identify and describe the ethical issues involved.

References

American College of Clinical Engineering (ACCE) (2011) Clinical Engineer (defined) [Internet] [Cited 3 Apr 2012]. http://www.accenet.org/default.asp?page=about&section=definition
Channel News Asia (2009) MOH alerts hospitals after two women given wrong dosage of cancer drugs [Internet] [Cited 2 Apr 2012]. http://www.channelnewsasia.com/stories/singaporelocal-news/view/1017828/1/.html
Darr K (2005) Ethic in health services management, 4th edn. Health Profession, Baltimore, MD
Deurenberg-Yap M, Foo LL, Low YY, Chan SP, Vijaya K, Lee M (2005) The Singaporean response to the SARS outbreak: knowledge sufficiency versus public trust. Health Promot Int 20:320–326
Dyro J (2004) Clinical engineering handbook. Elsevier, Amsterdam, pp 590–592
ECRI Institute (1992) Loose screws in TV mounts. Health Devices 21(11):427
Loughlin S, Williams JS (2011) The top 10 medical device challenges. Biomed Instrum Technol 45:98–104
The Free Dictionary (2012) [Internet] [Cited 4 Apr 2012]. http://medical-dictionary.thefreedictionary.com/Code+of+Ethics

Chapter 4
Ethics of Biomaterials for Implants

Dennis Kwok-Wing Tam and Oliver Faust

Keywords Biomaterials • Unnecessary operations • Professional conduct • Dental implants • Amalgam toxicity • Breast augmentation

Abbreviations

BME	Biomedical Engineering
PIP	Poly Implants Prosthesis
PMMA	Polymethyl Methacrylate
UBA	Umweltbundesamt
TCR	Tooth Coloured Restorations
WHO	World Health Organization

Introduction

Biomaterials are used to replace a missing biological structure, support a damaged biological structure, or enhance an existing biological structure (Basu et al. 2009; Park and Lakes 2007). The purpose of this reconstruction is to relieve pain, to restore function, and to facilitate healing (Martz et al. 1997). In most cases the biomaterial is shaped into a medical implant which is surgically placed inside the human body (Spiekermann 1995). Within the human body, the implant must interact with biological structures and systems (Williams 2009).

D. Kwok-Wing Tam (✉) • O. Faust
Electronic and Computer Engineering Division, School of Engineering,
Ngee Ann Polytechnic, 535 Clementi Road, Singapore 599489, Singapore
e-mail: tkd@np.edu.sg

J.Y.A. Foo et al., *Ethics for Biomedical Engineers*,
DOI 10.1007/978-1-4614-6913-1_4, © Springer Science+Business Media New York 2013

The use of biomaterials did not become practical until Lister had developed aseptic surgical techniques in 1860 (Lidwell 1987). Earlier surgical procedures, whether they involved biomaterials or not, were generally unsuccessful as a result of infection (van de Belt et al. 2001; Lidgren 2001). Problems of infection tend to be exacerbated in the presence of biomaterials, since the implant can provide a region which is inaccessible to the body's immunologically competent cells (Nelson and Williams 2007). The earliest successful implants, as well as a large fraction of the modern ones, were applied to the skeletal system. Around 1,900 bone plates were introduced to aid the fixation of fractures. Unfortunately, many of these early plates broke as a result of unsophisticated mechanical design: they were too thin and they had stress-concentrating corners (Pruitt and Chakravartula 2011). It was also discovered that materials, which were chosen for good mechanical properties, such as vanadium steel, corroded rapidly in the body (Manivasagam et al. 1997). Better designs and materials were soon introduced to overcome these problems. Following the introduction of stainless steels and cobalt–chromium alloys in the 1930s, greater success was achieved in fracture fixation. These developments paved the way for the first joint replacement surgery (Narayan 2009). Apart from metals and alloys, polymers play a key role in biomedical implants. The positive properties of this material were discovered by detecting warplane fragments within the bodies of pilots flying in World War II. These pilots were injured by pieces of Polymethyl Methacrylate (PMMA) plastic, a transparent material used as a substitute for glass in the aircraft canopy. The remarkable thing was that they did not suffer from human adverse chronic reactions which are normally caused by foreign material in the body. After this discovery, PMMA became widely used for corneal replacement and for replacements of damaged skull bone sections. Today, PMMA is used as bioactive filler in bone cement (Hamizah et al. 2012; Boger et al. 2008). Following further advances in materials and in surgical techniques, blood vessel replacements were tried in the 1950s (Chlupàc et al. 2009) and heart valve replacements (Deverall et al. 1985) and cemented joint replacements (Saha and Pal 1984; Vince et al. 1989) in the 1960s. Recent years have seen many further advances in infection prevention, material science, and implant design.

In this book chapter we adopt the position that there are two special areas, in the field of biomaterials for implants, which give rise to ethical concerns. One area is unnecessary operations, such as plastic surgery with implants, which might even be harmful to the patients body. In these operations biomedical implants replace or enhance non-vital body parts. Ethical issues arise from the fact that these operations are performed by medical professionals which were predominately trained by state run medical schools; therefore, the society should have a say to what purpose this medical skill is used. Furthermore, these operations might fuel vanity within the people having this treatment and envy within people who want that treatment, but for some reason cannot get it. Another area for biomedical implants, which gives rise to ethical issues, is the fact implants are manufactured. This manufacturing is not done by medical practitioners; it is done by commercial entities. These companies are set up to make money, and this primary requirement clashes with the primary requirement of the heath profession, namely to reduce suffering. Hence, there must be some sort of compromise that medicates between these distinct primary requirements. Both the outcome of this compromise and the way in which the compromise is obtained can become an ethical issue.

This chapter is organized such that the subsequent content supports our position. Section "An Overview of the Ethical Issues" provides a brief literature review on ethical issues in the wider field of biomedical implants. In the "Case Studies" section we put forward two case studies and discuss them. Each of the case studies is analyzed in terms of the ethical principles of beneficence, non-maleficence, autonomy, and justice. The "Conclusion" section provides concluding remarks and sets our work on ethics of biomaterials for implants into perspective with other medical and social issues.

An Overview of the Ethical Issues

Fundamentally, the most a transplant can achieve for a patient is the restoration of normal function. However, technological devices, such as biomedical implants, have the capacity to improve function to above-normal levels, because these devices are bound to scientific rather than evolutionary progress. Implant ethics therefore has to deal with issues of normality and disease, and with the admissibility of human enhancement. These questions do not arise in organ transplantation, because the transplant will always have about the same capability as the organ or tissue they replace. Furthermore, with current medical procedures, transplanted organs do not work well in the acceptor body.

It is predicted that future technological development will blur the distinction between transplantation and implantation (Hansson 2005). One concept which drives the merging of transplantation and implantation is "tissue engineering" (Lanza et al. 2007). This technique allows us to build organs, such as heart or liver, by growing cells on a technological scaffold that defines the structure required to produce an organ (Wood et al. 2003). The ethical issues, that dominate the debate on tissue engineering, are using human embryonic stem cells and therapeutic cloning (de Vries et al. 2008). For example, the ethical aspects of the use of stored tissue samples collected from minors are of topical interest (Trommelmans et al. 2009). However, there are other ethical questions, especially the ones which are concerned with the animal human relationship. One of these questions is: how do we measure the benefit to humans in relation to the disadvantages to animals? To answer this question we have to define the suffering of animals, the use of animals as means to an end, and the limited adequacy of the animal models (Oerlemans et al. 2010).

End of life decisions which concern donation are among the most pressing issues in transplantation ethics (Daly 2006). In implantation ethics, this problem does not exist; however, end of life decisions may reappear on the recipient's side. Implanted organs, in particular heart assist devices and heart replacements, can be life sustaining in the same way as external apparatus such as respirators, so the same type of end of life issues can be raised for these implants as for external devices (Ruark and Raffin 1988).

Distributive issues arise in health care whenever access to interventions is restricted, due to either natural limitations (as for organ transplantation) or to budget restraints (as may become the case for many implant devices). In the 1980s, research on artificial hearts was the target of critical discussion on the use of medical resources. It was argued that research on mechanical hearts should be stopped because the eventual cost of their deployment would be unbearable (Miles et al. 1990). More recently, it has been pointed out that advances in neuroscience, such as brain implants, have the potential to create and to remedy social inequalities. Therapeutic uses can reduce social inequalities, whereas enhancement implants available only to those who can pay for them would have the opposite effect (Roskies 2002; Maguire and McGee 1999; Parens 1995). Although distributive issues are essential for some implantation therapies, the severity of the problem is determined by the price of the intervention rather than whether or not it involves an implantation. In general, biomedical implants are mass produced and the production processes are well controlled through industry standard methodologies. Therefore, the artificial implant price is low, when compared to natural transplants. Hence, the only ethical distribution issues that arise come from the fact that even the relatively lower price might be too high for disadvantaged patients.

Mental function can be substantially influenced by implanted devices (Mason 1995) or by cell transplantation into the brain. This causes difficult problems that relate to mental change and personal identity. For example: Should the cognitive abilities of patients with dementia be improved at the price of changing their personality to such an extent that they are not perceived as the same people anymore? Recent advances in understanding and technology have led to the development of new treatments for brain diseases as well as defects of cognitive and behavioral functions (Chan and Harris 2006). An ethical problem arises when these methods are not only applied towards restoring brain function, as in the case of disease, but to enhance cognitive function for healthy individuals. The issues with such cognitive enhancement are: whether brain-enhancing treatments should be developed and made available and to whom; and what potential consequences might arise?

One form of implantation, namely cochlear implants, has been heavily criticized by members of the Deaf community for undermining their very existence (Tucker 1998). The community argues that the concept of a Deaf culture is growing. Hence, curing most hearing impairments means to extinguish a cultural movement.

Finally, fears have been expressed that nonvoluntary interventions may be carried out, perhaps in the form of brain implants used to control other human beings. Certainly, technical possibilities of manipulation through implantation are not far away. Electrical stimulation of a happiness center could make people addicted to this procedure, and other types of stimulation could change their perceptions of reality and perhaps make them easier to control (Altmann 2001). However, other much simpler means of manipulating and controlling people are already available. In the absence of a social setting in which someone seems to have a need for a brain implant in order to achieve the control he or she desires over other people, this does not seem to be an imminent danger.

The text below describes ethical issues that arise from the application of biomedical implants. We focus on practical issues which arise from the patient doctor relationship, and we discuss ethical decision making while working as a professional biomedical engineer. The practical aspect of these discussions is reflected in the fact that we have chosen case studies as method to deliver the argument.

Case Studies

Ethics of Professional Advice on Dental Implants

There is an increasingly popular trend for patients with dental problems to opt for tooth-colored materials, such as composite resins, in restorative dentistry. As a consequence, less patients choose silver amalgam, which is an alloy of mercury with one or more other metals. Apart from the fact that composite fillings appear to be more esthetically pleasing, the general public also fears individual exposure to mercury or mercury alloys. For a long period of time, it has been recognized that certain forms of mercury and its compounds show toxicological characteristics. Excessive intake of mercury may produce adverse health effects, such as bronchitis, pneumonia, renal damage in addition to disorder symptoms of the human central nervous system.

The following case scenario introduces some of the problems that arise from the fact that there are two methods for dental restoration. Based on this case study, we explore the ethics of professional advice on dental implants.

A Case Scenario

A man, who has amalgam fillings in most of his posterior teeth that lasted for 19 years, visited his dentist for scaling and polishing. His dentist detected some decay in one of the posterior teeth that had not been filled before. Upon discussing treatment plans for this new problem, the dentist suggested that the patient changes his amalgam restorations to Tooth Coloured Restorations (TCRs). He cited that TCR would be esthetically pleasing and toxin free, unlike amalgam. The patient then proceeded to have all his amalgam restorations changed to TCR. Later, he found out that TCRs are less durable and reliant than amalgam restorations. He was also dismayed to find out that, in the process of replacing his amalgam restorations, some healthy tooth structures were removed.

Facts

This case study explores the controversial use of dental restoration methods. Before we embark on an ethical analysis, it is necessary to examine the benefits and drawbacks of amalgam as a biomaterial for dental restoration and under what circumstances it is safe to use. Amalgam has been used in dental restorations for over 150 years, and it remains an effective restorative material. It may be considered as the material of choice for some restoration in posterior teeth from the perspectives of longevity (Maguire and McGee 1999, p. 53).

Most dental amalgams are called *silver amalgams* since silver is the principal constituent that reacts with mercury. There are several types of silver amalgams. A conventional dental amalgam alloy contains between 67 % and 74 % silver, with 25–28 % tin, and up to 6 % copper, 2 % zinc, and 3 % mercury (this little amount of mercury is used to facilitate the amalgamation reaction). The amalgam alloys are mixed with mercury before clinical placement at a 1:1 weight ratio. The mercury content of a finished dental amalgam restoration is therefore approximately 50 % by weight (Roskies 2002, p. 16).

Mercury is one of the few metallic elements that is liquid at room temperature. As a consequence, it is able to undergo an alloying reaction with other elements at ambient temperatures to form, in a clinically acceptable time (solidify in a few minutes and gradually harden over a few hours), a customized mass that can be adapted to the size and shape of a tooth cavity. Furthermore, it is strong enough to resist the forces of occlusion for many years (Sutow et al. 2007).

To assess the toxicity of amalgam, it is important to note that there are several different forms of mercury. First there is elemental mercury itself, a volatile form of liquid metal, referred to as Hg0. Second, mercury is stable in two other oxidation states (Hg^{1+} and Hg^{2+}) and it is able to form inorganic compounds, of either monovalent or divalent form. Third, mercury is able to form a variety of organic compounds, including methylmercury. There is a clear connectivity between these forms with respect to the global cycle of mercury (Nielsen et al. 2006). Each form of mercury has its own toxicological profile, although, in general terms, the toxicity of these forms is highest with the organic mercury compounds, followed by elemental mercury and inorganic mercury compounds. This is important when considering different exposure routes to these forms.

The indications for mercury exposure are normally obtained by measuring mercury levels in urine and blood of individuals. Autopsy/postmortem studies give an indication of the lifetime mercury exposure of individuals, during their lifetime, which includes exposure to dental amalgam. As such, these studies suffer certain unquantifiable limitations. Therefore, data dealing with blood and urine mercury concentration are generally considered to be more relevant, because they reflect actual exposure.

Mercury is distributed ubiquitously in the environment; therefore, it can be taken up via food, water, and air. Dietary intake is the most important source of nonoccupational exposure to methylmercury, with fish and other seafood products being the dominant source. Intake of elemental mercury from dental amalgams is another source contributing to the total mercury burden in humans (WHO 1990, 1991). The provisional tolerable weekly intake has been established at 1.6 μg/kg of body weight. Because the two major sources of mercury body burden include dietary intake of methylmercury and intake of elemental mercury from dental amalgams, mercury is inevitably present at low concentrations in human tissues. Mercury has been detected in blood, urine, human milk, and hair in individuals in the general population. The mercury concentrations in whole blood of individuals with or without amalgam fillings are usually below 5 μg/l blood, but these concentrations do depend on dietary habits and the number of amalgam fillings (ATSDR 1999; BAT 1997).

Intake Estimates for Mercury from Dental Amalgams

Mercury vapor is released from silver amalgam restorations during chewing, tooth brushing, and parafunctional activities including bruxism. The parameters of this release of mercury vapor by amalgam depends on the number of fillings, the filling size and placement, chewing habits, food texture, grinding and brushing teeth, nose–mouth breathing ratio, inhalation–absorption, ingestion and body weight, and the surface, composition, and age of the amalgam restorations. Therefore, there are large variations in the estimation of daily mercury absorption and release.

Mercury, released from dental amalgam, distributes in the oral cavity as inhalable mercury vapor, or is dissolved in saliva after oxidation or suspended in it as amalgam particles. With respect to systemic exposure assessment, only the inhaled fraction is relevant, since elemental mercury and inorganic mercury are poorly absorbed from the GI-tract; therefore, they have only a minor contribution to systemic exposure. The daily uptake of mercury from amalgam fillings is estimated to be up to 27 μg/day in individuals with a large number of fillings. One study shows an intake from 1 to 5 μg/day from dental amalgam for people with seven to ten fillings. The World Health Organization (WHO) reported a consensus average estimate of 10 μg/day of amalgam derived mercury (range: 3–17 μg/day) (WHO 1991). Weiner and Nylander (1995) estimated the average uptake of mercury from amalgam fillings in Swedish subjects to be within the range of 4–19 μg/day (Nylander and Weiner 1991). Skare and Engqvist (1994) estimated that the systemic uptake of mercury from amalgams in middle-aged Swedish individuals with a moderate amalgam load (30 surfaces) was, on average, 12 μg/day (Skare and Engqvist 1994).

The mercury body burden of dental personnel is normally higher than in the general population. The mean urine mercury levels in dental personnel has been variously reported to range from 3 to 22 μg/l, compared to 1–5 μg/l as the normal range for nonoccupational groups (Hörsted-Bindslev 2004). This increased body burden is attributed to dental personnel mixing and applying dental amalgam and removing amalgam restorations; Ritchie et al. (2004) showed that dentists had, on average, urinary mercury levels over four times that of control subjects although all but one dentist had urinary mercury levels below the UK Health and Safety Executive health guidance value (Ritchie et al. 2004). Dentists were significantly more likely than control subjects to have suffered from disorders of the kidney; the researchers concluded that these symptoms were not significantly associated with their level of mercury exposure.

Correlations have been found among dentists between urinary mercury levels and the number of hours worked in the surgery ($r = 0.22$, $P = 0.006$) and the number of amalgam restorations placed ($r=0.38$, $P<0.001$) and removed ($r=0.29$, $P<0.001$) in a week, with urine mercury levels in dentists ranging from 0.02 to 20.90 (mean 2.58) nmol mercury per nmol creatinine. A confounding factor in such investigations is the number of amalgam surfaces dentists have in their own mouths (Ritchie et al. 2002).

Toxicity of Elemental Mercury

Due to the very low absorption of elemental mercury after oral intake, we focus on the toxic effects observed after inhalation of elemental mercury. There are a number of reviews that address the toxicity of elemental mercury (MAK Kommission der Deutschen Forschungsgemeinschaft [DFG] 1999; BAT Kommission der Deutschen Forschungsgemeinschaft [DFG] 1997; United Nations Environment Programme 2002; ATSDR [Agency for Toxic Substances Disease Registry] 1999; IRIS 2002). The assessment of elemental mercury toxicity is mainly based on observations in occupationally exposed humans. Inhalation of extremely high concentrations of elemental mercury, in excess of 10 mg/m^3, may produce bronchitis and pneumonia, in addition to symptoms of the central nervous system. However, such concentrations are many orders of magnitude above those encountered through the release of elemental mercury from dental fillings. Early signs of toxicity after inhalation of mercury are less specific and the early phase of toxicity is often referred to as "micromercurialism." Clinical findings in this condition are tremor, enlargement of the thyroid, increased uptake iodine in the thyroid, tachycardia, gingivitis, and hematological changes. To diagnose the early stage of elemental mercury intoxications, at least three of these findings should be present along with increased mercury concentrations in blood or increased mercury excretion with urine.

Quantitative data on elemental mercury inhalation exposure, mercury concentrations in blood and urine, and early effects of mercury toxicity have been established. The nonspecific symptoms of micromercurialism are observed at long-term exposures to elemental mercury air concentrations of 0.05 mg/m^3, or at concentrations of mercury of 35 μg/l in blood or 150 μg/l in urine. Overt neurotoxicity (tremor) occurs after long-term inhalation of elemental mercury at concentrations between 0.1 and 0.2 mg/m^3 with resulting blood mercury concentrations between 70 and 140 μg/l and urinary mercury in the range of 300–600 μg/l (MAK Kommission der Deutschen Forschungsgemeinschaft [DFG] 1999; BAT Kommission der Deutschen Forschungsgemeinschaft [DFG] 1997; United Nations Environment Programme 2002).

Due to the small dose received by inhalation of mercury from amalgams, a direct comparison of maximal mercury air concentration in the oral cavity of individuals with amalgam fillings and occupational limits for air concentrations of mercury requires consideration of absorbed dose. The inhalation of mercury at the occupational exposure limit results in an uptake of more than 300 μg of Hg per day, whereas inhalation of mercury from dental amalgams gives body burdens which are at least 20-fold lower than those resulting from occupational exposures at present limits for air concentrations (ATSDR [Agency for Toxic Substances Disease Registry] 1999; IRIS 2002).

Based on the evaluation of several longitudinal studies involving blood samples to determine mercury content over a prolonged time period, the German MAK-Commission (tasked to set occupational exposure limits that are without health risks) concluded that even many years of mercury exposure to concentrations that result in urinary mercury levels of 100 μg/l or even higher do not cause objective adverse effects (MAK Kommission der Deutschen Forschungsgemeinschaft [DFG] 1999). The urinary mercury levels were equivalent to mercury concentrations in blood of approximately 23 μg/l. The BAT-value (maximal permissible concentration of hazardous compounds or their metabolites in body fluids) was therefore set at 100 μg/l of urine or 25 μg/l of blood and is considered a no-adverse-effect-concentration for mercury in humans (BAT Kommission der Deutschen Forschungsgemeinschaft [DFG] 1997). For the general population, the Federal Environment Agency (Umweltbundesamt [UBA]) derived reference values including general background exposure to mercury from various sources (fish and seafood consumption, mercury in other foods) of 1.4 μg/l of urine and of 2 μg mercury/l of blood in adults without amalgam fillings and with low seafood consumption. According to UBA, no adverse effects of mercury are observed at blood levels lower than 5 μg/l (including pregnant women) and urinary mercury concentrations lower than 0.7 μg/l. These assessments included both inorganic mercury and the more toxic methylmercury (UBA [Umweltbundesamt] 1999).

Analysis

There is no scientific evidence that any of the alloys, which are currently used in dental amalgam, constitute a risk of adverse health effects in individuals apart from allergic reactions to the individual elements (Roskies 2002, p. 25). Moreover, dental amalgams were found to be acceptably safe to use (Roskies 2002).

We emphasize that dental amalgam remains an effective restorative material and, from the perspectives of durability, reliability, and economic performance, may be considered the material of choice for some restorations in posterior teeth. However, there are established disadvantages of dental amalgam (a) it is not tooth colored and (b) it does not adhere to remaining tooth tissues. Therefore, its use has been decreasing in recent years and tooth-colored filling materials have become increasingly popular, consistent with the general trend towards more minimal intervention techniques in dentistry. There has been, for some years, a move towards non-amalgam, adhesive, TCR. This trend shows some variations within and between countries, and is emphasized by the significant reduction of training in the placement of dental amalgam restorations and the corresponding increase in training in the use of amalgam alternatives in a growing number of dental schools. It is accepted that the reducing the amount of mercury used for human activity would be beneficial, both for the general decrease in human exposure and from environmental considerations.

However, with respect to the debate about the possibility of causal relationships between the use of mercury containing amalgam and a wide variety of adverse systemic health effects and taking into account many studies and investigations into this putative causal link, there is no unequivocal evidence to support this possibility. However, it is generally concluded that no increased risks on adverse systemic effects exist, and we do not consider that the current use of dental amalgam poses a risk of systemic disease. It is recognized that some local adverse effects are occasionally seen with dental amalgam fillings, but the incidence is low and normally readily managed. It is also recognized that there have been reports of reactions to dental amalgam, which are not supported by scientific evidence, but indicate that very occasionally an individual may have unexplained atypical physical or other reactions attributed to mercury.

Ethical Principles to Consider

- *Beneficence*: In this ethical analysis we assume that the advice given by the dentist was based on the believe that composites fillings are more esthetically pleasing, and hence they are more suitable for his patient. This assumption is based on the fact that the case study does not provide sufficient information on this topic. By giving the benefit of a doubt to the dentist, we consider that his actions were caused by a lack of knowledge and not by profit considerations.
- *Non-maleficence*: By advising the patient to undergo dental surgery, which, according to the facts stated in case study, is unnecessary the dentist violates the patient's right of non-maleficence by potentially subjecting him to an unnecessary operation.
- *Autonomy*: The dentist may be acting on the spirit of beneficence when he suggested a replacement to TCR as he firmly believes that it is better. However, by not providing the due facts of TCR properties, he infringes on the patient's autonomy. By not receiving the adequate information needed, the patient's autonomy was compromised.
- *Justice*: From the case scenario we recognize that both patient and dentist lack sufficient knowledge of the facts stated in case study. However, the patient aims to address this lack of knowledge by consulting a dentist. The dentist is a member of the medical profession specializing in dental care. Therefore, the patient had reason to believe that a dentist could help by providing an unbiased account of amalgam versus TCR. In terms of justice, the lack of knowledge for a medical professional weighs more than the lack of knowledge for the layman patient. Especially, when the patient was aware of his lack of knowledge, and the lack of knowledge of the dentist was only uncovered through wrong advice.

Ethics of Professional Conduct in Biomedical Engineering

There are psychological and esthetical reasons for women wanting breast augmentation. Psychological reasons include increased confidence and self-esteem. Esthetic reasons come from the need to restore the original body shape after disease-related breast removal. Unfortunately, who seek breast augmentation women could be the victims of unethical biomedical companies who supply them with substandard implants. Biomedical professionals who are working in the production line, quality assurance, or marketing areas of these companies should have an understanding of their responsibility/duty of upholding ethical principles. As an example, we discuss ethical issues, related to a hypothetical case scenario of a biomedical engineering (BME) engineer working in a biomedical company that produces implants for plastic surgery. The case study is set against the background of newspaper reports on the alleged wrongdoing of the Poly Implants Prosthesis (PIP) company.

Case Scenario

A BME engineer has been working at PIP before this news has been released. He was in charge of monitoring both production and quality of breast implants. Instructions were given from the company management that, in order to bring down cost, industrial instead of medical grade silicone would be used as implant filler. This decision was justified with a use case scenario. To be specific, both filler materials are silicone; hence, they are essentially the same. The only difference between them is that medical grade silicone is more bio-inert[1] than industrial grade silicone. However, the human body does not come into contact with the potentially harmful filler material, because it is kept within an outer layer of bio-inert material.

Facts

This case study explores a crisis that arises due to somebody's malpractice. In this case many parties are being affected as a results of greed, negligence, and lack of regulatory control. The question of who should be responsible and bear the financial burden of replacing over 30,000 unsafe implants, which were supplied by PIP to women in Europe who had undergone the breast augmentation.

Before we start the ethical analysis, we have to gather information on the breast augmentation business in Europa. In particular, we are concerned with the question: which materials are used to produce breast implants?

Breast augmentation is a popular cosmetic surgery in Europe. In France, there are over 400,000 women who have breast implants (Guardian Web page). Around 21,000 breast augmentations are carried out each year in that country. In the UK, there is also a significant increase in the demand for breast augmentation surgery among women. The UK plastic surgery statistics, supplied by the British Association of Aesthetic Plastic Surgeons, shows that 10,015 people had this type of surgery in 2011 and numbers rose by 6.2 % from 2010.

Breast implants are biocompatible materials which were designed by biomedical engineers for cosmetic surgery. Medical-grade silicone is commonly used as filler

[1]Does not influence the function of the human body.

in breast implants. Medical grade silicones have to undergo a series of tests before they are characterized as being safe for human implantation. These silicones have a reduced content of low molecular weight polymers. Therefore, this type of silicone reduces the risk of patients to develop adverse reactions, such as rashes and, in more severe cases, cancer, when compared to industry standard silicone. However, the increased risk, of not using medical grade silicones, does not stop unethical biomedical implant manufacturers to use cheaper materials with lower purity levels, which could endanger the health of patients, just for the sake of higher profits.

Here is a summary of the news reported, in December 2011, about a case of "poisonous" breast implants, which surfaced in many European countries (Guardian web page).

A company, PIP, was found to have been cutting corners and saving over SGD$1.7 billion a year by using industrial grade silicone instead of medical-grade silicone as filler in their breast implants. This cost cutting decision allowed PIP to offer the breast implants at a very competitive price. When the news about PIP's scandal surfaced, many women, who had breast augmentation, experienced high levels of psychological stress. It has also been reported that there are known cases of cancer in women who had PIP implants.

Both French and British authorities state that regular tests for ruptures and leakages should be done by those who have PIP implants. In response to news, UK health-care regulator said in 2011 that no evidence to show the filling in PIP implants being chemically toxic or carcinogenic. On the contrary, some French doctors call for women to err on the side of caution to have even their non-ruptured breast implants removed.

Full details of the case can be found: http://www.guardian.co.uk/lifeandstyle/2011/dec/14/france-faulty-breast-implant-scandal.

Analysis

Breast implant surgery is rather common, due to esthetic or psychologic reasons. Hence, the repercussions of implant toxicity can be far-reaching, and they have devastating effects on woman who have undergone this type of surgery.

Among those woman who are seeking breast implants, there are two main groups: those whose reason is esthetics and those who have undergone cancer treatment and require breast reconstruction. There are different responses to the plight of victims based on their reasons for undergoing breast implant surgery. Socially, victims may be demeaned or ridiculed and they have a publicity that they did not desire. Inevitably, their privacy would be infringed, because of PIP's malpractice because they have to disclose the fact that they had surgery in order to appeal for compensation. Furthermore, there are unforeseen financial issues for these victims to address and cope with.

The question of who should be responsible for the outcome is raised. This is a controversial issue that involves the victims, PIP, healthcare providers, and governing authorities. There are hefty compensations that may come in the form of monetary expense, a marred image for PIP and healthcare providers as well as loss of time spent because of law suits. Furthermore, the process of establishing responsibilities is stressful as well as psychologically and emotionally draining for the victims involved.

Healthcare providers (surgeons and hospital procurement staff) may be called to conduct more stringent checks on their implant suppliers. Being medically trained and informed, they are trusted to be in a position to validate the safety of products used. The call of some surgeons from France for victims to remove their implants (even if it has not ruptured) is an example of them providing medical advice for the interests of patients.

This case also highlights the need for regulatory parties to conduct customary checks on the quality control. It takes a concerted effort, not just by individuals, but also various organizations on a large scale to ensure the safety standard of body implants.

The public also deserves to know how this issue will be addressed and what remedial action will be taken to solve it. To allay fears, it is paramount for governing authorities to step forward and account for what has happened and what they would do to ensure this cannot happen again.

As a BME professional in PIP, there are a number of ethical issues to consider. When confronted with the information that PIP is using nonmedical grade silicone, he will have to make a decision of whether or not to inform the authorities about the misconduct of the company management.

Continuing with the job and observing what is really going on in the company is another option a BME may take. This wait and see approach allows further appraisal of the situation and observation of the effects which are caused by the company's decision. He still has the option of reporting to the authorities at a later date.

Alternatively, an indifferent perspective may be taken, where the impact of using nonmedical grade silicone filler is of no concern to the BME professional. The absence of a sense of responsibility and careless about the implication of the company's decision allows him to continue work as per normal.

Lastly, the BME professional may choose to take himself out of the situation by leaving the company. This can be motivated also by personal reasons or the realization that as an individual there is nothing he can do to change the decision of the company.

One should be aware that there are extrinsic factors that affect the wider environment of the health-care world (e.g., government policies, negligence, company policies, company malpractice, etc.). Hence, it is important to realize that there may be multiple parties involved in this ethical issue.

Ethical Principles to Consider

- *Beneficence*: PIP did not act in the interest or welfare of its customers when they intentionally used nonmedical grade silicone fillers. This was a breach of public trust and its consumers by acting inhumanely and negligence of its duty of care to them. A BME professional, who abides by the company's stance, would also not be acting in beneficence to the victims of the potentially toxic implants.

- *Non-maleficence*: The BME engineer believes that no harm is done during the normal operation of the breast implant. The change from medical grade to industrial grade silicone filler has only harmful effects for patients when the implants rupture or leak. Therefore, he has the burden to weigh up the probability of failure and harm which is done when such failures occur.
- *Autonomy*: Patient's autonomy has been compromised when PIP concealed the truth about the type of silicone filler used in their products. There was no full disclosure between the two parties. This evidently affected the decision-making process of the patient.
- *Justice*: To discuss justice we have to realize that there is a conflict between patient and PIP. The patient has the feeling of being treated unfair, hence the harmonious relationship has been disturbed. Justice is concerned with reestablishing this harmonious relationship. Before this can be done, it is essential to establish the extend of harm done by PIP. In this case, we find that there is reversible and irreversible harm. To be specific, the bill from PIP for the breast implant, which was paid by the patient, can be considered as a reversible harm done to the patient, because this transaction can be reversed. However, it is impossible to reverse the pain felt by the patient during the operation. Hence, the only civilized option is to establish the monetary equivalent for the pain, i.e., monetary compensation. After having established that the harmonious relationship of patient and BME company is disturbed and harm was done, we have to find a basis for a claim. This brings us into the domain of legal considerations. The patient, or the lawyers of the patient, needs to establish that violation of law occurred or that the claimant (patient) is entitled to a legal remedy. This legal claim can be based on the fact that the decision to use industrial grade instead of medical grade silicone was based on just a use case. Not considering fault cases, such as rupturing of the outer shell of the implant, was grossly negligent. This judgment is based on the general agreement that both companies who build human implant have to have high safety standards where use and fault cases need to be considered.

Conclusion

The ethics of professional advice is very important for medicine, because in most cases the person seeking medical advice is no expert and the advice itself has direct consequences for the health of the patient. Therefore, the first of the two case studies explores ethics of professional advice on dental implants. The person seeking advice does so because he or she expects a benefit, in this case curing dental disease, correcting tooth alignment, or esthetic improvements. The dentist should give advice as unbiased as possible to maximize the benefits for the patient. However, there are financial and organizational issues which might lead into ethical conflicts. These ethical conflicts are also present for professional conduct in biomedical engineering. The case study on the fabrication of breast implants goes to the heart of the conflict between financial gain and reducing suffering. The limitation for safety is always cost, and it is an ethical question on how much money is spent on safety. The engineering question is how this money is translated into an increased safety level. However, it is undoubtedly incorrect to provide wrong or misleading information. Therefore, regulation of the biomedical implants is of paramount importance. We address the ethical issues of regulation with a case study.

In this book chapter we adopt the position that there are two special areas, in the field of biomaterials for implants, which give rise to ethical concerns. Advice on biomedical implants requires expertise in both medical and biomechanical fields. Ethical issues arise from the fact that advice has always financial and health consequences. So, biomedical implants are a complex topic and advice on this matter is most of the time biased. Therefore, ethical issues arise in terms of beneficence, nonmaleficence, autonomy, and justice. Another area for biomedical implants, which gives rise ethical issues, is the fact implants are manufactured. This manufacturing is not done by medical practitioners; it is done by commercial entities. These companies are set up to make money, and this primary requirement clashes with the primary requirement of the heath profession, namely to reduce suffering. Hence, there must be some sort of compromise that medicates between these distinct primary requirements. Both the outcome of this compromise and the way in which the compromise is obtained can become an ethical issue.

These ethical issues on biomedical implants are here to stay for a long time, because they are a direct consequence of current industrial and medical systems. The core of the conflict is a clash between the humanistic principle of eliminating pain and suffering with the capitalist principle of making capital gain. The best thing to do is to construct a benefiting scenario. To be specific, both humanistic and capitalistic principles pose different and sometimes conflicting requirements on the biomedical implant. However, these differences and conflicts can be resolved trough discussion. It is our firm believe that ethical principles should be used to guide this discussion.

References

Altmann J (2001) Military uses of microsystem technologies. Dangers and preventive arms control. Agenda Verlag, Münster

ATSDR (Agency for Toxic Substances Disease Registry). (1999). Toxicological profile for mercury—update. Atlanta-GA. http://www.atsdr.cdc.gov/toxprofiles/tp46.html. Accessed 11 Dec 2012

Basu B, Katti DS, Kumar A (2009) Advanced biomaterials: fundamentals, processing, and applications [hardcover]. Wiley, Hoboken, NJ

BAT Kommission der Deutschen Forschungsgemeinschaft (DFG) (1997) Mercury, metallic mercury and inorganic mercury compounds. In: Triebig G, Schaller KH (eds) Analyses of hazardous substances in biological material, vol 3. Wiley-VCH, München, pp 123–142

Boger A, Bohner M, Heini P, Schwieger K, Schneider E (2008) Performance of vertebral cancellous bone augmented with compliant pmma under dynamic loads. Acta Biomater 4(6):1688–1693

Chan S, Harris J (2006) Cognitive regeneration or enhancement: the ethical issues. Regen Med 1:361–366

Chlupàc J, Filová E, Bacáková L (2009) Blood vessel replacement: 50 years of development and tissue engineering paradigms in vascular surgery. Physiol Res 58(Suppl 2):S119–S139
Daly BJ (2006) End-of-life decision making, organ donation, and critical care nurses. Crit Care Nurse 26(2):78–86
de Vries RBM, Oerlemans A, Trommelmans L, Dierickx K, Gordijn B (2008) Ethical aspects of tissue engineering: a review. Tissue Eng Part B Rev 14:367–375
Deverall PB, Campalani G, Anderson DR (1985) Heart valve replacement. Arch Dis Child 60:1111–1112
Hamizah AS, Mariatti M, Othman R, Kawashita M, Noor Hayati AR (2012) Mechanical and thermal properties of polymethylmethacrylate bone cement composites incorporated with hydroxyapatite and glass-ceramic fillers. J Appl Polym Sci 125(S1):E661–E669
Hansson SO (2005) Implant ethics. J Med Ethics 31(9):519–525
Hörsted-Bindslev P (2004) Amalgam toxicity: environmental and occupational hazards. J Dent 32(5):359–365
IRIS (2002) Methylmercury. In: Integrated risk information system. Database quest, last revised: 12/03/2002. US-EPA
Lanza R, Langer R, Vacanti JP (2007) Principles of tissue engineering. Tissue engineering intelligence unit. Elsevier Science, Burlington, MA
Lidgren L (2001) Joint prosthetic infections: a success story. Acta Orthop Scand 72(6):553–556
Lidwell OM (1987) Joseph Lister and infection from the air. Epidemiol Infect 99:569–578
Maguire GQ, McGee EM (1999) Implantable brain chips? Time for debate. Hast Cent Rep 29(1):7–13
MAK Kommission der Deutschen Forschungsgemeinschaft (DFG) (1999) Mercury and inorganic mercury compounds. In: Greim H (ed) Occupational toxicants—critical data evaluation for mak values and classification of carcinogens by the commission for the investigation of health hazards of chemical compounds in the work area, vol 15. Wiley-VCH, München, pp 81–122
Manivasagam G, Dhinasekaran D, Rajamanickam A (1997) Biomedical implants: corrosion and its prevention—a review. American Society for Testing and Materials
Martz EO, Goel VK, Pope MH, Park JB (1997) Materials and design of spinal implants—a review. J Biomed Mater Res 38(3):267–288
Mason RO (1995) Applying ethics to information technology issues. Commun ACM 38:55–57
Miles SH, Siegler M, Schiedermayer DL, Lantos JD, La-Puma J (1990) The total artificial heart: an ethics perspective on current clinical research and deployment. Health Policy 16(2):175–176d
Narayan R (2009) Biomedical materials. Wiley
Nelson KE, Williams CM (2007) Infectious disease epidemiology: theory and practice, 2nd edn. Jones and Bartlett Publishers, Sudbury, MA
Nielsen E, Ladefoged O, Larsen JC (2006) Risk assessment of contaminant intake from traditional Greenland food items. DFVF publikation, Department of Toxicology and Risk Assessment, Danish Institute for Food and Veterinary Research, Denmark
Nylander M, Weiner J (1991) Mercury and selenium concentrations and their interrelations in organs from dental staff and the general population. Br J Ind Med 48(11):729–734
Oerlemans AJM, Rodrigues CHCML, Verkerk MA, van den Berg PP, Dekkers WJM (2010) Ethical aspects of soft tissue engineering for congenital birth defects in children—what do experts in the field say? Tissue Eng Part B Rev 17:229–234
Parens E (1995) The goodness of fragility: on the prospect of genetic technologies aimed at the enhancement of human capacities. Kennedy Inst Ethics J 5(2):141–53
Park JB, Lakes RS (2007) Biomaterials: an introduction. Springer, New York, NY
Pruitt LA, Chakravartula AM (2011) Mechanics of biomaterials: fundamental principles for implant design. Cambridge texts in biomedical engineering. Cambridge University Press, Cambridge, UK
Ritchie KA, Gilmour WH, Macdonald EB, Burke FJT, McGowan DA, Dale IM, Hammersley R, Hamilton RM, Binnie V, Collington D (2002) Health and neuropsychological functioning of dentists exposed to mercury. Occup Environ Med 59(5):287–293

Ritchie KA, Burke FJT, Gilmour WH, Macdonald EB, Dale IM, Hamilton RM, McGowan DA, Binnie V, Collington D, Hammersley R (2004) Mercury vapour levels in dental practices and body mercury levels of dentists and controls. Br Dent J 197(10):625–632

Roskies AL (2002) Neuroethics for the new millennium. Neuron 35(1):21–23

Ruark JE, Raffin TA (1988) Initiating and withdrawing life support. N Engl J Med 318(1):25–30

Saha S, Pal S (1984) Mechanical properties of bone cement: a review. J Biomed Mater Res 18(4):435–462

Skare I, Engqvist A (1994) Human exposure to mercury and silver released from dental amalgam restorations. Arch Environ Health 49(5):384–394

Spiekermann H (1995) Implantology. In: Rateitschak KH, Wolf HF (eds) Color atlas of dental medicine. G. Thieme Verlag, New York, NY

Sutow EJ, Maillet WA, Taylor JC, Hall GC, Millar M (2007) Time-dependent corrosion potential of newly-placed admixed dental amalgam restorations. Dent Mater 23:644–647

Trommelmans L, Selling J, Dierickx K (2009) An exploratory survey on the views of European tissue engineers concerning the ethical issues of tissue engineering research. Tissue Eng Part B Rev 15(3):241–247

Tucker BP (1998) Deaf culture, cochlear implants, and elective disability. Hast Cent Rep 28(4): 6–14

UBA (Umweltbundesamt) (1999) Kommission, "Human-biomonitoring des umweltbundesamtes," Berlin: Stoffmonographie Quecksilber—Referenz-und Human-Biomonitoring-(HBM)-Werte. Bundesgesundheitsbl. Gesundheitsforsch Gesundheitsschutz 42:522–532

United Nations Environment Programme (2002) Governing Council, UNEP/Chemicals—Global mercury assessment. United Nations Environment Programme

van de Belt H, Neut D, Schenk W, Horn JRv, van der Mei HC, Busscher HJ (2001) Infection of orthopedic implants and the use of antibiotic-loaded bone cements: a review. Acta Orthop Scand 72(6):557–571

Vince KG, Insall JN, Kelly MA (1989) The total condylar prosthesis 10- to 12-year results of a cemented knee replacement. J Bone Joint Surg Br 71(5):793–797

WHO (World Health Organisation) (1990). Environmental Health Criteria 101, Methylmercury. Geneva: World Health Organisation, International Programme on Chemical Safety.

WHO (World Health Organisation) (1991). Environmental Health Criteria 118, Inorganic mercury. Geneva: World Health Organisation, International Programme on Chemical Safety.

Weiner JA, Nylander M (1995). An estimation of the uptake of mercury from amalgam fillings based on urinary excretion of mercury in Swedish subjects. Sci Total Environ 168:255–265

Williams DF (2009) On the nature of biomaterials. Biomaterials 30(30):5897–5909

Wood S, Jones R, Geldart A (2003) The social and economic challenges of nanotechnology. Technikfolgenabschätzung 12(3):72–73

Chapter 5
Ethics and Data Mining in Biomedical Engineering

Andrew P. Bradley

Keywords Machine learning • Pattern recognition • Confidentiality and privacy • Unbiased estimates • Performance measures • Receiver operating characteristic (ROC) curves • Statistical evaluation

Introduction

Data mining is a generic term given to the process of analysing data, usually high volumes of data contained in large databases, in order to discover previously unknown patterns and trends. Data mining utilizes and combines methods from statistics, machine learning, pattern recognition and database management. Typical data mining tasks involve detecting data subsets that are similar in some way, unusual or anomalous or have features that are associated or dependent. Although data mining is not traditionally focused on the development of predictive models that generalise known patterns to new (unseen) data, this is often an extremely valuable way of verifying the efficacy of the derived models. That is, if these models accurately predict unseen data, then it is more likely that these models truly represent the underlying patterns in the data rather than being purely by-chance occurrences. In addition, in biomedical applications the purpose of the data mining is often to better understand the patterns of disease so that improved diagnoses, prognoses and treatments can be developed in the future.

Like most statistical methodologies data mining, in itself, is ethically neutral (Seltzer 2005). However, there is a clear public expectation that data mining, especially in Biomedical Engineering, will be undertaken ethically. Such research relies on public trust that researchers will behave ethically and this trust adds to the

A.P. Bradley (✉)
School of Information Technology and Electrical Engineering, The University of Queensland, St. Lucia, Brisbane, QLD 4072, Australia
e-mail: a.bradley@itee.uq.edu.au

J.Y.A. Foo et al., *Ethics for Biomedical Engineers*,
DOI 10.1007/978-1-4614-6913-1_5, © Springer Science+Business Media New York 2013

Engineer or Clinician's ethical responsibilities. In addition, people frequently participate in research studies for altruistic reasons, primarily for the benefit of others. Ethics, therefore, plays an important role in protecting participants and needs to be explicitly considered when planning research projects, acquiring data and designing data mining algorithms. In addition, there are clear expectations, policies and guidelines from many of the professional societies whose members work in areas directly related to data mining. For example:

- The American Statistical Association's Ethical Guidelines for Statistical Practice (http://www.amstat.org/about/ethicalguidelines.cfm)
- The Association for Computing Machinery's Code of Ethics and Professional Conduct (http://www.acm.org/about/code-of-ethics)
- The Institute of Electrical and Electronic Engineers' code of ethics (http://www.ieee.org/portal/pages/iportals/aboutus/ethics/code.html)
- The International Association for Pattern Recognition's statement of ethics (http://www.iapr.org/constitution/soe.php)

In addition, institutions where research involving human and animal participants is undertaken, such as universities and hospitals, have themselves a responsibility to promote the responsible and ethical conduct of research. These institutions will typically have a number of specialist ethical review committees whose responsibility is to support and facilitate research where ethical principles and the well being of research participants are paramount. However, before approaching an ethics committee, it is necessary to fully consider all of the ethical issues involved in the research and to design your experimental protocol accordingly. Research in Biomedical Engineering may often require ethical review from all collaborating universities, hospitals and organisations. However, it is often the case that once one institution has acknowledged ethical approval, then it can be expedited at other institutions.

The design of experimental methodologies requires making many ethical and technical decisions. When making decisions, one of the most important issues for any person is that they must be able to justify the decisions that they have made, showing that they did the best they could in the circumstances (Preston 1996). This includes both how and why they made their decisions. At the very least they must be respectful of the rights and views of all participants, adequately consider all short- and long-term imperatives and minimise all negative effects and risks.

Reproducibility is one of the primary principles of the scientific method and increasingly for research in general. The production of reproducible research is not only good practice but also increases its impact (Vandewalle et al. 2009). However, reproducible research may not always be ethical, but it is undoubtedly ethical to present research so that someone can accurately and independently reproduce it.

It is important to note that there is a clear difference between unethical and illegal behaviour. Ethics is concerned doing what is right, good and fair. It is what we ought to do, rather than what is legal, most convenient or acceptable in the circumstances. Obeying the law is mandatory, but as individuals we have the ability to make ethical choices, behave ethically and uphold ethical principles. Experience shows that when research goes wrong, or an individual or organisation behaves unethically, then there is a public demand for the law to be changed to stop this from recurring. Therefore, both legal and ethical expectations change, and hopefully improve, over time. To this end, this chapter is written with the aim of better informing researchers on the ethical considerations when applying data mining techniques to Biomedical data.

Ethics and Data Mining

Ethical issues relate to all research involving human or animal participation in its many forms. Experimental research can take many forms, but in the context of data mining it typically relates to the acquisition and analysis of data relating to the participants of a research study. The primary aims of these experiments should at least be one of the following: to improve knowledge, diagnosis or treatment of disease; to provide commercial, individual or societal benefits or to enable the design, development or validation of new devices, materials or techniques. Therefore, every data mining experiment should not only have a clear aim, that is justifiable to a layperson based on potential benefits, but should also be designed using appropriate methods, based on the current literature and utilising appropriate techniques. In addition, the experiment should be conducted and supervised by suitably qualified, competent and experienced staff and at all times should respect the rights, beliefs and expectations of all participants.

It should be noted that while data mining experiments are often undertaken on clinical data, and in a clinical setting, they are not formally "clinical trials". The term clinical trial is reserved for projects that trial the efficacy of a drug, device, therapy, intervention or treatment as part of a formal regulatory approval process, such as for the Food and Drug Administration (FDA).

Traditionally, the ethical issues involved in a data mining experiment can be separated into three main areas (Seltzer 2005):

- Overall aims
- Confidentiality and privacy
- Suitability and validity of methods employed

I will discuss and elaborate on these issues in the following sections. However, the focus of this chapter will be on recommending unbiased statistical methodologies that properly and ethically evaluate the suitability and validity of data mining algorithms for specific applications in Biomedical Engineering.

Overall Aims

The overall aims of a data mining experiment should be explained in terms that:

- Describe *what* the purpose and goals of the research project are
- Justifies *why* the research project is being undertaken, outlining the problem it is addressing and its potential benefits and outcomes

- Detail *how* the research will be undertaken and in particular what participation in the research project involves

Clearly, from the above, the overall aims of any data mining research experiment are critically important. However, these aims should not be considered in isolation and should always be addressed via an appropriate experimental methodology. In practice, this means that a risk-benefit analysis must be explicitly performed to directly demonstrate that the ends justify the means. That is, that the potential benefits of the project outweigh any potential harm associated with the actual experiment. As an example, the benefits of a post hoc analysis of clinical data should outweigh any potential risks, say that of a participant being identified from the data. Clearly, there must be significant benefits for any experiment that involves exposing the participants to anything other than a low or negligible risk, i.e. where the foreseeable risks are more than discomfort or inconvenience (risks can also be related to the foreseeable added risks above the risks of everyday living). In addition, all experiments (even low risk ones) should be designed so as to minimise and mitigate all risks by reducing either the severity of harm or its likelihood of occurrence (or both). In addition to the explicit risks associated with the experimental design, specific protection must also be afforded to vulnerable or other special populations who may be subject to special risks or who may not be fully able to protect their own interests. Examples of such special populations are children, indigenous peoples, mentally impaired or perhaps the elderly.

The overall aims of the project should provide sufficient information to put all ethical considerations related to the research into context. It should be written in everyday, layman's language that is suitable for inclusion in a participant information sheet and can be understood and appreciated by non-expert readers, including potential participants. Details of participation procedures should be specified, including the expected duration, location and frequency of participation.

Data mining algorithms can perform a variety of tasks that can be either fully or semi-automated. Typically, algorithms search large databases to find records that are similar to a specified query example, e.g. given a set of symptoms return similar cases and their prognoses or return a list of associated symptoms, related diagnoses and conditions. However, from an experimental viewpoint these tasks can be converted, and hence simplified, into dichotomous (binary) classification decisions. For example, the "find similar cases" example can be defined as successful (a true positive) if a truly similar case is returned in the top, say 10, records from the database. Alternatively, a grouping or clustering algorithm can be judged successful (true positive) if a known test sample is associated with the correct subgroup. Alternatively, a sample from another subgroup would be considered a true negative if it were not associated with the positive subgroup. In this way, the algorithms can be evaluated based on simple metrics such as classification accuracy or error rate. The methodologies for reliably estimating these metrics are explained in more detail on the section on suitability and validity of methods.

Sample size determination is necessary to determine the number of participants or measurements required for a statistically meaningful research experiment to be undertaken. Sample size is often a critical feature of any empirical research study in which the goal is to make an inference about a population using just a sample. In practice, the sample size used in a study is determined based on the time, expense and risks associated with data collection and the need to have sufficient statistical power so that one can draw meaningful conclusions from the experiment (Zar 1998). Ethical considerations imply that we must determine a "goldilocks" sample size that is large enough to be statistically meaningful, but not so large as to unnecessarily cause inconvenience to participants or expose them to unnecessary risks.

Estimating the required sample size to adequately develop and evaluate a data mining algorithm is not trivial. In a typical data set there are complex interrelationships between a number of domain specific unknowns. For example, either prior knowledge or some form of exploratory data analysis must be used to estimate the intrinsic dimensionality of the data (the number of features required to classify the data); the complexity of the problem (the type of decision boundary, classifier type or parameter settings appropriate for the data) and the expected performance (the level of performance we can hope to achieve on this data). As already discussed, often the objective in data mining is to retrieve specific cases from a large database, which is representative of an underlying population. In this case, we can estimate the number of samples required to demonstrate one of two things:

1. That the observed performance has not been obtained by a purely random labelling of the test examples.
2. That the observed performance obtained using one method (say, feature set, classifier or parameter setting) is superior to another.

Both of these situations require the explicit formulation of a *null hypothesis* that expresses the concept of "no difference". That is, that the performance is equal to a random labelling or that two methods do have equal performance. As will be elaborated on shortly, the measure of "performance" here is usually the mean value of error rate estimated over a set of data. Therefore, our hypothesis test becomes: that the estimated (observed) mean is equal to a specific value (say, 0.5) or that the difference between two means is zero (Zar 1998). Of course, if it is concluded that the null hypothesis is false, then we can reject this hypothesis and accept that an *alternate* hypothesis is true. It is often this alternate hypothesis that is of most practical interest as it allows us to conclude that a data mining algorithm performs better than random labelling or that one method is superior to another. Examples using a number of methods of sample size estimation, in the context of rare event detection, are presented in Bradley and Longstaff (2004). A worked example is also presented at the end of this chapter.

On occasion data mining experiments are based on data acquired from participants who are undertaking a standard diagnostic or treatment protocol. When this is the case it is vitally important that results of the experiment do not affect the ongoing clinical process. The aim of the experiment is to design or validate data analysis algorithms. Therefore, by definition, these algorithms are experimental and have not (yet) demonstrated their efficacy. Therefore, the results of such preclinical trials are for research purposes only. Clearly, if the data acquisition and data analysis are separate processes and the data is properly de-indentified, then this issue can be avoided.

Confidentiality and Privacy

A longstanding regulation in the research arena is that individuals must give "informed consent" regarding their participation in a research experiment. This consent must cover not only the actual information or data they are providing but also all possible future uses intended by the facility receiving the data and performing the experiment. This clearly highlights that individuals must be seen as active participants in a research experiment, rather than mere subjects of the experiment. The term informed consent also highlights the importance of the material given to the participants in order to solicit their participation (typically a participant information sheet), in that it must be written in clear, plain, non-technical language. The potential risks and benefits of the research must also be clearly communicated. This is particularly important when the informed consent is given by a third party, such as the parent of a child participant. In any case, it is often recommended to have an independent witness sign the informed consent form.

Specifically, it is recommended that participants are not only given an explicit assurance of confidentiality, but are made aware of the following, before any data is collected:

- The purpose of the data collection and any data mining projects that utilise this data
- How the data will be used
- Who will have access to and be able to analyse the data
- Access to any results of the data analysis
- The security surrounding access to the data
- How collected data can be updated
- How participation can be withdrawn without penalty
- How participants can obtain more information, report adverse events, problems or concerns

For data mining experiments requiring access to patient records, a gatekeeper, or permission giver, is required. The role of a gatekeeper is to apply specified and agreed criteria to allow passage of clinical data to the researcher for experimental analysis. Most importantly the gatekeeper must not have a conflict of interest and take responsibility for their decisions. This person is authorised to write a letter on behalf of an organisation involved with the research, such as a clinic or hospital, which gives permission to the researcher to access the population of potential participants under their authority. For example, if you wish to conduct research in a hospital and the participants are the clinicians, then gatekeeper approval will need to be obtained from the relevant hospital authority and the clinical director or appropriate manager before you may approach those clinicians for recruitment.

Data that has been made anonymous must not be aggregated with other data so that it can subsequently be identified (Seltzer 2005). In addition, researchers must ensure that all data, particularly data containing personal information (that is, information that can identify the person), are secure both at the point of storage and during transit. For example, data must be stored in a locked filing cabinet, computer hard-drive protected by password, encryption or de-identification of data. Researchers must also be aware of relevant legislation and guidelines governing privacy.

Data in the public domain typically does not require ethical approval for analysis, unless the participants can be identified from the data. However, data gathered in public places, such as surveys and observations, may still require ethical approval, even though individual informed consent may never be sought. Again, one of the key criteria here is whether the "participants" could be identified from the data and whether they had an expectation of privacy. In addition, a clinician must not give a researcher a list of patients who may be potential participants for a research study unless those patients have previously authorised disclosure of their information to non-clinical parties.

Another issue that is often overlooked, when designing research experiments, is the disclosure of potentially valuable intellectual property (IP). For example, if researchers have developed a novel, non-obvious, data mining algorithm with clinical applicability, then they may wish to file a provisional patent on this invention. However, if they use this algorithm as part of a clinical trial prior to submitting a provisional patent, then the patent examiner may rule that the invention is no longer novel as it had prior commercial or public use (Old 1993). There are two potential solutions to this issue: first, a non-disclosure agreement could be integrated into the paperwork that participants agree to and sign, second, the experiment could be redesigned so that the novel (inventive) parts are only applied in-house (under commercial confidence) and not in public. This may be a preferable solution if the data acquisition stage of the experiment is not novel and can be separated from the novel data mining stage of the experiment.

Suitability and Validity

As pointed out by Friedman (1995), no one data classification method is universally superior to any other one over all possible problem data sets. More recently this has become known as the "no free lunch" theorem (Duda et al. 2001). From this theorem, it is clear that each data mining or classification algorithm potentially has a class of target functions or data sets for which it is best suited. Put another way, given a data set, it is important to find the classification algorithm that is best suited to this data. Therefore, data mining experiments can be seen as an attempt to investigate which algorithm should be used on a particular problem data set or subset of problems. For example Bradley (1997) considered the subset of "medical diagnostic" problems characterised by six public domain data sets. The conclusions drawn from such an experiment are, therefore, targeted towards a specific subset or class of problems and should not be extrapolated beyond the scope of that experiment.

When designing an experiment there are a number of forms of bias that can potentially affect the results, thus potentially leading to incorrect conclusions being drawn. Specifically, a biased experimental design can produce (apparently) superior results for one particular algorithm and inferior results for another. Therefore, we must be aware of and explicitly minimise all potential forms of bias:

- *Selection bias* can occur from the inappropriate selection of the specific data set or participants for analysis that favour one algorithm over another.
- *Expert bias* can occur when the data mining expert analysing the data spends more time and effort developing and tuning one particular algorithm than the others. Expert bias can be minimised by not attempting to tune any of the algorithms to the specific data set and wherever possible using default values of algorithm parameters. These parameters may include the pruning parameters for decision trees, the value of *k* for a nearest neighbour algorithm or the learning rate, momentum and initial conditions for a neural network. This naive approach may result in lower estimates of performance, but it is a bias that should affect all of the algorithms equally. Alternatively, one can attempt to tune the performance of each algorithm on the data set, but then differing levels of expertise with each method may still be advantageous to some algorithms. Additionally, experimentation time increases dramatically as you evaluated different input representations, input transformations, network architectures, learning parameters, pruning parameters or attempt to identify outlier samples in the training set.

Of course, as we have already discussed, particular algorithms are naturally biased towards certain types of data. However, this bias is the focus of the experiment and while important, would not normally be minimised.

Experimental Methodology

Before proceeding to discuss specific experimental methodologies, it is worthwhile to briefly define some basic terminology:

- *Data set*: A number of labelled instances of a problem or concept, each instance consisting of a number of features or measurements (from one upward) and its class label. The classification and measurements are usually assumed to be complete and true. That is, labelled with the "gold standard" classification, although this need not always be the case in practice. In addition, not all samples need be individually labelled with the advent of semi-supervised and multiple instance learning methods, such as (Dietterich et al. 1997).
- *Training set*: A subset of the data set that is used to train or estimate the parameters of the algorithm used to classify previously unseen instances in the test set.
- *Test set*: The subset of the data set not contained in the training set that is used to test the classification algorithm and estimate a measure of performance, such as the error rate.
- *Error rate*: There are two error rates commonly used (Weiss and Kulikowski 1991).
  - The re-substitution or *apparent error rate*, which is defined as the error rate of the classification algorithm based on the data used to train the algorithm, i.e. via a re-substitution of the training data.

– Estimates of the *true error rate*, that is, given an asymptotically large number of examples the classification algorithm has never seen before, what is the probability that the algorithm will not predict their class correctly.

Both of these error rates are measured in terms of

$$\text{error\_rate} = \frac{\text{number\_of\_errors}}{\text{number\_of\_instances}}.$$

The apparent error rate is not typically a useful measure of classifier performance. This is because it does not predict the generalisation ability of the classifier on the future, previously unseen, examples. The performance of the system on examples it has not seen before is a very useful indication of the system's potential performance in practice. Unfortunately, the true error rate of a classifier cannot be calculated, except under the assumption of Normal (Gaussian) class conditional probability density functions (Hand 1981). Also, error rate is made difficult to measure due to the asymptotically large numbers of testing examples that are often required. Therefore, typically only *estimates* of the true error rate can be obtained. These estimators are referred to as "biased" if they estimate the error rate to be either too low (optimistic bias) or too high (pessimistic bias). Clearly, the apparent error is an optimistically biased estimator of the true error rate; it converges to the true error rate only with an infinite training set. For example, in the case of a rote learning system, such a nearest neighbour algorithm, the apparent error rate will always be zero (as the "test" sample can always be found, correctly labelled, in the training set). However, it can be shown that the true error rate of a rote learner only approaches zero as the size of the training set approaches infinity (Duda et al. 2001).

Error rate estimators also have a *variance* associated with them, which defines the confidence limits to which the true error rate has been determined. Generally speaking, the more computationally intensive an estimator is, the less bias it will have and the better we can measure its variance. There are a number of techniques for estimating the true error rate, brief details of which are given below. They are presented approximately in order of computational complexity:

- *Single Train and Test or holdout*. Here the data set, containing n instances, is split into one training set of size $(n-j)$ and one test set of size j. The usual proportions of the data are 2/3 in the training set and 1/3 in the test set. The error rate on the test set is then used to estimate the true error rate. The holdout estimate is usually pessimistically biased and no estimate of the variance of the error rate is obtained. The variance of the error rate is an important indication of how sensitive the error rate is to variations of instances in the train and test sets, i.e. if we ran the holdout test again how would the error rate vary? The holdout estimate of the true error rate converges to the true error rate when there are over ~1,000 examples in the test set (Weiss and Kulikowski 1991).
- *Cross-validation*. For data sets, numbering perhaps thousands of examples, a single train and test methodology will give a good estimate of the true error rate.

However, for most practical classification problems there is a severe limitation on the amount of data available, and so more complex approaches must be taken to estimating the true error rate. One of the most popular approaches is a random sub-sampling technique called cross-validation. General v-fold cross-validation consists of randomly partitioning the data set in v subsets, $(v-1)$ of these subsets are then used to train the classifier, whilst the remaining subset is used to test the classifier. This process is repeated v times, rotating the subset used to test the classifier and training on the remaining $(v-1)$ partitions. In this way, each instance in the data set is used to test the classifier once, while still having the majority ($v-1$ partitions) of the data available to train the classifier. The estimate of the true error rate is then the average error rate of each of the v test subsets. The cross-validation sampling technique is random, but should ensure that the approximate proportions of samples of each class remains constant in each subset. This slight adjustment to maintain the prevalence of each class does not bias the error estimates and is supported in the research literature (Breiman et al. 1984). The extreme case of cross-validation is called *leave-one-out*. Leave-one-out is actually n-fold cross-validation, where n is the number of instances in the data set. Here a classifier is generated using $(n-1)$ cases and then tested on the single remaining case. This process is repeated n times and the estimate of the true error rate is the number of errors divided by n. Cross-validation provides an almost unbiased estimate of the error rate, though it can have a large variance depending on the number of partitions, v, chosen, and the data set size. This problem can be reduced by performing repeated (re-sampled) cross-validation and averaging the resultant error rates.

- *Bootstrap*. The Bootstrap is a sample with replacement technique, so that when instances are drawn from the data set to be included in the training set, they are replaced back into the data set. This means that examples can appear more than once in the training set. The 0.632 bootstrap is so called because the probability that an example will be selected for the training set from a data set of size n, drawing n samples, is 0.632. Therefore, there will be, on average, 63.2 % unique samples in the training set. The test set is then constructed from the remaining (un-sampled) 36.8 % of the data set. This procedure is repeated approximately 200 times and the error rate estimates on the test set averaged to give e0. The Bootstrap estimate, 0.632B, is then the linear combination of:

$$0.632\text{B} = 0.368 \times \text{app} + 0.632 \times \text{e0},$$

where app is the apparent error on the whole data set. Both e0 and 0.632B are low variance estimates of the true error rate, though e0 is usually pessimistically biased. On larger data sets the 0.632B estimate can be optimistically biased, but it works well on small data sets depending on the true error rate (Weiss and Kulikowski 1991).

- *Jackknife*. The sampling methodology of the jackknife estimate is exactly the same as that outlined for the leave-one-out estimate. The generalised jackknife being the same as v-fold cross-validation. However, the jackknife attempts to

Table 5.1 Holdout, cross-validation and Bootstrap estimators of the true error rate

	Holdout	v-fold cross-validation	Leave-one-out	Bootstrap
Train set	j	$n-(n/v)$	$n-1$	n (j unique)
Test set	$n-j$	n/v	1	$n-j$
Iterations	1	v	n	~200

estimate the bias of the apparent error rather than estimate the true error directly using a test set. The jackknife estimate is an estimate of the bias of the apparent error rate, as the sample size, n, goes to infinity (McLachlan 1992, Sect. 10.2). The jackknife estimate, therefore, produces a different (though related) estimate of the true error rate than that produced by cross-validation. The true error rate is calculated as the apparent error rate plus a bias term. This bias term is calculated using a linear approximation that assumes the apparent error rate with an infinite sample size that is an estimator of the true error rate (Efron 1982, Chap. 2). The error rate estimated using the jackknife is a linear approximation of the bootstrap; hence the bias is dependent on sample size (Efron 1982, Chap. 7).

Table 5.1 shows how the computational complexity increases from the holdout error estimate to the Bootstrap estimate and how the number of training examples increases from around 2/3 of the data set to nearly the whole data set. The main advantage of these sub-sampling techniques is that all of the data is used for testing at least once during the iterations.

Another important point to make here is that the sample size, n, above refers to the number of *independent* samples. Therefore, in experiments where multiple measurements or instances are acquired from each participant, the sample size is the number of participants, *not* the total number of instances. This is because it is often difficult to argue that samples acquired from the same participant are independent. For example, in each fold of a leave-one-out cross-validation, the classifier should be trained on the data acquired from all of the participants, *except* the "test" participant's data.

The question still remains, however, of which estimator to use on a particular data set. The following guidelines are based on extensive Monte Carlo simulations, detailed in Weiss and Kulikowski (1991, Chap. 2):

- If the sample size >2,000 use a holdout method. This is based on a Binomial trial with 1,000 randomly drawn test samples used to estimate the true error rate; it is independent of the actual class distributions.
- If the sample size >100 use tenfold cross-validation or leave-one-out. The tenfold cross-validation method is much less computationally intensive than leave-one-out and can be used with confidence with data sets numbering in the hundreds. The classification and regression tree (CART) procedure (Breiman et al. 1984) was extensively tested with varying numbers of cross-validation partitions, and tenfold cross-validation was found to give a good trade-off between accuracy and computational complexity.

Table 5.2 Confusion matrix

	Predicted class		
True class	Negative	Positive	
Negative	TN	FP	CN
Positive	FN	TP	CP
	RN	RP	*N*

- If the sample size <100 use leave-one-out. The leave-one-out estimate is an unbiased estimate of the error rate; however, sufficient samples are required to reduce its variance.
- If the sample size <50 use leave-one-out and in addition use the 0.632 Bootstrap and twofold cross-validation repeated 100 times. The leave-one-out estimate should be used except when:
  - The leave-one-out estimate is less than the 0.632B estimate, when the 0.632 estimate should be used.
  - The leave-one-out estimate is greater than the repeated twofold cross-validation estimate, when the repeated twofold estimate should be used.

Performance Measures

The previous section focused on the estimated classification accuracy and its complement error rate. However, in biomedical applications error rate is often not a meaningful performance metric as prior probabilities can be highly skewed, e.g. the incidence rates for many diseases are significantly lower than 5 % in the natural population. In these cases a by-chance classifier can obtain accuracies greater than 95 % simply by labelling instances randomly, based on the prior probabilities, or labelling all instances as normal (negative).

For binary classification problems there are two types of errors, usually called *false positives* and *false negatives*. A false positive is a classification of positive given to a test instance that is actually negative, and a false negative is the negative classification of a test instance that is actually positive. Here the terms positive and negative are abstract, but are interpreted to mean: "has the condition of interest" and "does not have the condition of interest", respectively. Therefore, the classifications made by a *dichotomizer* on a test set can be summarised as the number of true positives, TP, true negatives, TN, false positives, FP and false negatives, FN. This is shown in Table 5.2 and is known as a confusion matrix. A confusion matrix is a form of contingency table showing the differences between the true and predicted classes for a set of labelled samples.

In Table 5.2, CN and CP are the numbers of true negative and positive examples in the data set; RN and RP are the number of predicted negative and positive samples and *N* the total number of samples.

Although the confusion matrix shows all of the information about the classifier's performance, more meaningful measures can be extracted from the confusion matrix to estimate relevant performance criteria, for example:

- *Accuracy* (1 − *error*) = (TP + TN)/(CP + CN). This estimates the probability of correctly labelling a test sample, by combining results from both classes in proportion to the class priors.
- *Sensitivity* = TP/CP. This estimates the probability of correctly detecting a positive test sample and is independent of class priors. It is also known as *recall* rate (Landgrebe et al. 2006).
- *Specificity* = TN/CN. This is the complement to sensitivity and indicates the probability of correctly detecting a negative test sample invariant of the class priors.
- *Positive predictive value* (PPV) = TP/RP. This indicates the fraction of the positive samples detected that are correctly labelled. PPV estimates an overall posterior probability for a model and is, therefore, a meaningful performance measure when detecting rare events. However, PPV combines both results from both positive and negative samples and so is dependent on the class priors. It is also known as *precision* (Landgrebe et al. 2006).
- *Negative predictive value* (NPV) = TN/RN. This is the complement to PPV. It is also known as *purity*.
- *Posfrac* = (TP + FP)/*N*. This measure is useful in applications requiring second-stage manual processing of the positive outcomes of the classifier, such as medical screening tests, and estimates the reduction in manual effort provided by the classification model.

For a particular decision threshold, say with known priors and equal misclassification costs, the average and standard deviation of these performance metrics can be calculated over, say the ten cross-validation test partitions. These estimates are of course not necessarily appropriate when the Binomial approximation of the Normal distribution does not hold, but they do give an indication of the central tendency and variability of the classifier's performance.

The Receiver Operating Characteristic Curve

In many empirical evaluations it is common to assume that the prior probabilities of each class are known and fixed. A further assumption often made is that the respective misclassification costs are known, allowing for the optimal decision threshold to be found (Fukunaga 1990). Here, performance measures such as error rate may be applied to compare different models as appropriate. However, in biomedical applications, misclassification costs cannot be specified exactly, and class priors may not be reflected by the samples in the data set, or even worse, the priors may vary. For example, the experimental data set acquired may have been "enriched" with far more positive cases than would be seen in the underlying population. Additionally, the prior probability of a disease such as cervical cancer will vary dependent on whether the participant was recruited from a general practice or sexual health clinic. Consequently, optimal threshold selection in biomedical applications is an ill-defined problem and so algorithm evaluation based on a fixed threshold is usually inadequate.

Receiver Operator Characteristic (ROC) curve analysis has become a useful and well-studied tool for the evaluation of classifiers (Fawcett 2006), especially in biomedical applications (Bradley 1997). Measures such as the area under the ROC (AUC) allow for a performance evaluation independent of costs and priors by integrating performance over a range of decision thresholds. This can then be viewed as a performance measure that is integrated over a region of possible operating points, priors and misclassification costs. In addition, this approach can also be extended to multi-class decisions provided the number of discrete is not too large (Landgrebe and Duin 2008; Swets et al. 2000)

A ROC curve can be constructed from the set of confusion matrices obtained by varying a classifier's decision threshold, e.g. by varying misclassification costs or class priors. The ROC curve is then the plot of the true positive rate, TPR, versus the false positive rate, FPR, where:

- TPR = TP/(TP + FN)
- FPR = FP/(TN + FP)

Traditionally, the ROC curve was used to measure how well a receiver can detect signal from background noise, but in general it measures the ability of a classifier to distinguish positive from negative instances.

In statistical pattern recognition this method of treating the two types of errors separately is called the *Neyman–Pearson* criterion (Fukunaga 1990). Here we fix one of the class error probabilities, usually from a performance specification, and then minimise the other class error probability with this constraint. Because of the difficulties in solving this constrained minimization, a practical approach to the Neyman–Pearson method is to vary the decision threshold between two class distributions and plot the locus of the points obtained on a ROC curve.

There are, of course, a number of ways in which a multidimensional discriminant function can be adjusted to produce a ROC curve, e.g. with rotations as well as translations of the decision surface. However, it is normal to assume that the form of the discriminant function does not change at different points on its ROC curve, i.e. that none of the parameters of the fitted model change, just the threshold at which the positive/negative decision is made. In the case of a neural network, where the discrete output classes are obtained by thresholding the output neuron, the default (equal cost) decision threshold would be at the mid-point of the neuron's output range, e.g. at 0 for an output range of [−1, 1]. A ROC curve can then be produced by testing the neural network at a number of decision thresholds within the range of the classification neuron's output, e.g. [0, 0.1, 0.2,…, 0.8, 0.9, 1]. The connection strengths (weights) between the neurons in the network are then not changed, only the decision threshold is varied. A variety of classifiers such as nearest neighbour, decision trees and discriminant functions can be modified in this way to produce ROC curves (Bradley 1997).

The Area Under the ROC Curve

As a classifier's decision threshold is varied, each successive point on the ROC curve can be used to calculate the area under the ROC curve. This statistic, referred to as AUC, can be estimated in a number of ways, but one of the simplest and most easily applied to a variety of classifier types is trapezoidal integration. It is also possible to calculate AUC by assuming that the underlying probabilities of predicting negative or positive are Normally distributed. The ROC curve will then have an exponential form and can be fitted either: directly using an iterative Maximum Likelihood (ML) estimation, or if the ROC curve is plotted on double probability paper, a straight line can be fitted. The slope and intercept of this fitted line are then used to obtain an estimate of AUC.

As noted in Hanley and McNeil (1982), the trapezoidal approach systematically underestimates AUC. This is because of the way all of the points on the ROC curve are connected with straight lines rather than smooth concave curves. Therefore, it is important to minimise this bias by ensuring that an adequate number of points (say ~15) are plotted. The trapezoidal approach has the advantage of not relying on any assumptions as to the underlying distributions of the positive and negative samples and is exactly the same quantity measured using the Wilcoxon test of ranks (Hanley and McNeil 1982).

Typically AUC will be calculated for a classifier on each of the tenfold cross-validation partitions. In this way, the mean and standard deviation of the AUC can be estimated. It should be noted that there are two distinct possibilities when it comes to combining the ROC curves plotted from different test partitions (Swets and Pickets 1982, pp. 64–66):

- *Pooling*. Here, the raw data (TPR and FPR) are averaged so that one average (group) ROC curve is produced from the multiple estimates at each point on the curve. For the case of tenfold cross-validation we would have ten estimates of TPR and FPR for each point on the ROC curve. The assumption here is that each of the classifiers produced on each of the training partitions comes from the same underlying population. Although this assumption may be true in terms of their overall discrimination performance, the assumption that for each partition they are all estimating the same points on the ROC curve is less palatable. This can be seen from the fact that pooling the data in this way depresses the combined index of performance. For example, take two estimates of a point on a ROC curve, (0.03, 0.32) and (0.15, 0.44); the average of these two points is on a straight line midway between them. This average point is at (0.09, 0.38) and is an underestimate of the true average point that would lie on a concave curve connecting the two points, at about (0.08, 0.39). Pooling done in this way will, therefore, lead to an AUC with a pessimistic bias. Therefore, averaging is typically chosen in preference to pooling.

- *Averaging*. Here we average the actual accuracy index extracted from each of the ROC curves obtained from each train and test partition. In this way, AUC is calculated for the ROC curves and then averaged. This gives both an estimate of the true area and its standard deviation. The only problem with this approach is that it does not result in an average ROC curve, only an average AUC. Therefore, the pooled responses are typically used when we actually wish to visually observe the whole ROC curve.

The standard error of the AUC (Hanley and McNeil 1982) is of importance if we wish to test the significance of one classification scheme producing a higher AUC than another. Conventionally there have been three ways of calculating this variability associated the AUC (Hanley and McNeil 1983):

1. From the confidence interval associated with the maximum likelihood (ML) estimate of AUC
2. From the standard error of the Wilcoxon statistic, SE(W)
3. From an approximation to the Wilcoxon statistic that assumes the underlying positive and negative distributions are exponential in type (Hanley and McNeil 1982)

This final assumption has been shown to be conservative, i.e. it slightly overestimates the standard error, when compared to assuming a Gaussian-based ROC curve (as in the ML method).

Statistical Evaluation

Using an appropriate train and test methodology, a series of classifiers can be produced and their performance evaluated. However, it is unlikely that the performance of one classifier will be clearly superior to the others. Typically, there is a large degree of variation both between classifiers and for each classifier over the different train and test partitions. Therefore, we should attempt to draw an inference from this experiment by utilizing a statistical hypothesis test, such as an Analysis of Variance (ANOVA).

An ANOVA allows us to test the null hypothesis that a number of classifier's have, on the average, the same performance. If there is evidence to reject the null hypothesis, then we can test the alternative hypothesis that one of the classifiers has better performance than the others. ANOVA is simply an extension of Hypothesis tests of means (such as the *t* and *F* tests) to the case of multiple groups, which, in our case, is multiple classifiers (Zar 1998). ANOVA avoids the necessity of performing multiple hypothesis tests for each pair of classifiers as we effectively test all hypotheses simultaneously. The experimental design allows us to compare, on each data set, the mean performance for each learning algorithm and for each train and test partition.

Typically a two-way ANOVA would be applied that tests two hypotheses:

1. H_0', that all the means are equal due to the different train and test partitions
2. H_0'', that all the means are equal due to the different classifiers

Of these two hypotheses we are only really interested in the second, H_0'' and we could have used a one-way ANOVA to test this hypothesis alone. However, a one-way ANOVA assumes that all the groups are independent (which is unlikely seeing as they were trained on the same data) and can often be a less sensitive test than a two-way test that uses the train and test partitions as a blocking factor (Zar 1998). Conventionally an ANOVA utilises the F-test; however, other non-parametric tests can also be applied, for example the Friedman test utilises rank statistics.

When the ANOVA test on a specific performance measure produces evidence to reject the null hypotheses, then we can accept the alternative hypothesis that classifier performance differs. However, we still do not know which of the means are significantly different from which other means. Therefore, a post hoc (or follow-up) test is typically used to separate significantly different means into subsets of homogeneous means. Post hoc tests, such as Tukey's honestly significant difference criterion, typically compare every group mean with every other group mean and incorporate a mechanism for controlling errors where the null Hypothesis is incorrectly rejected (referred to as a Type I error). Alternatively, one can limit the number of multiple comparisons carried out, say if we are only interested in comparing one classifier to all of the others. Additionally, we can also apply a Bonferroni correction to compensate for the multiple (independent) comparisons. That is, by testing the alternative hypothesis at a level of significance that is divided by the number of comparisons made. However, this adjustment is conservative (Zar 1998).

Summary

In this chapter we have outlined why ethical considerations are important when performing data mining research on biomedical data. We have also made some detailed recommendations for practitioners who need to design ethical and valid data mining experiments covering the salient issues of overall aims, risk analysis, confidentiality, suitability and validity. In particular, we have detailed how an appropriate experimental methodology may be planned so that we can estimate an adequate sample size, efficiently partition training data so that we can train and then optimise our algorithms and finally select a suitable metric with which to evaluate the algorithm's performance on an independent test set. By following these guidelines a data mining researcher will be able to formulate a hypothesis and draw statistically meaningful conclusions from their research experiments.

Sample Size Estimation: Worked Example

The following example concerns the clinical evaluation of an algorithm that is used in a device for automatically testing the hearing of newborn infants. The data is collected on a conventional device and then processed again, off-line, using the proposed algorithm. The trial investigates two dependent variables: test time (the time taken for the algorithm to either pass or refer an infant) and the test accuracy. It explains the way in which the sample size is calculated and provides a justification of the sample size in terms of its statistical validity.

Test Times

This estimate is based on reports in the literature that the current devices take an average of approximately 4 min to complete a single assessment, for which we will assume a variance of approximately 4 min. Shown below are the approximate samples sizes that will be needed to identify the listed reductions in this test times using a paired *t*-test ($p<0.05$, $\sigma=4$ min) with a 90 % power value:

- Reduction from 4 to 3 min: 170 participants
- Reduction from 4 to 2 min: 44 participants
- Reduction from 4 to 1 min: 21 participants
- Reduction from 4 to ½ (<1) min: 16 participants

Based on initial test we believe the new algorithm will reduce the average test times to less than ½ min. Therefore, we will need less than 16 participants to be able to show that the new algorithm is faster than the current devices.

Test Accuracy

In the following we will use calculations involving acceptable failure rates (AFR) and the area under the receiver operating characteristic (ROC) curve (AUC), as per Bradley and Longstaff (2004).

Using the AFR approach with the assumption (obtained from the literature) that conventional hearing screening device has a sensitivity of 99 % and a specificity of 95 %, and a minimum detectable difference ($p<0.05$) of 2 % (i.e. we can reject the null hypothesis that a measured specificity equalled the reference specificity when the measured specificity was greater than or equal to 97 %), the required sample sizes are:

- 32,200 (31,878 passes + 322 refers) for a "natural" (1 %) prior probability of refer
- 1,074 (752 passes + 322 refers) for an "enriched" (30 %) prior probability of refer

Using the AUC approach with the assumption (obtained from the literature) that conventional hearing screening device has an AUC of 90 %, and a minimum detectable difference ($p<0.05$) of 2 % (i.e. we can reject the null hypothesis that a measured AUC equalled a reference AUC when the measured AUC was greater than or equal to 92 %), the required sample sizes are:

- 28,900 (28,611 passes + 289 refers) for a "natural" (1 %) prior probability of refer
- 1,044 (731 passes + 313 refers) for an "enriched" (30 %) prior probability of refer

Therefore, we predict that the trial will require approximately 1,200 participants in order to demonstrate that the proposed detection algorithm is more accurate than current device. In summary:

1. There is a requirement for 1,000 participants, based on calculations involving acceptable failure rates (AFR) and the area under the receiver operating characteristic (ROC) curve (AUC).
2. Another 200 participants are required to account for an expected data rejection rate of 20 %, i.e. an expectation that 20 % of the data will not be used in the final analysis because it is contaminated by user error, physiological artefact, electrical artefact, etc. This figure of 20 % is based on previous experience during pilot trials of a prototype device in a clinical setting.
3. The use of an "enriched" sample of participants who score a referral (a fail) on their conventional hearing assessment. This enrichment will raise the prior probabilities in the AFR and AUC calculations from 1 to 30 % and will significantly reduce the total number of participants required for the study. It will be achieved by recruiting into the research project all participants who score a "refer" result on their conventional hearing assessment.

Review Questions

1. Discuss, which of the following activities are most likely to require ethical review as being human research?
   (a) Surveys, interviews or focus groups
   (b) Undergoing physiological or medical testing or treatment
   (c) Being observed by researchers
   (d) Acquiring access to personal documents, medical records or other materials
   (e) Access to personal information as part of an existing published or unpublished database
2. Discuss which of the following example projects are likely to be considered for an expedited ethical review. In making your decision consider whether the projects are likely to involve low risk or negligible risk:
   (a) A research project is to be carried out on school children where they undertake a series of tests, including: questionnaires, picture card recognition and word recognition. The aim of the project is to study the effect of television on their development, with the possibility of undertaking the same tests in another 2 years time.

(b) A researcher wishes to observe the daily work practices of laboratory technicians and to ask them to complete an anonymous questionnaire about their work practices. The ethics application form is supported by a copy of the gatekeeper approval from the Laboratory manager, a copy of the questionnaire, including an information sheet for participants. The application also makes it clear that the observation involves note taking only.

3. Perform a risk analysis, considering all potential physical, psychological, sociological, economic and legal risks, for the following research experiments:

   (a) Surveys are mailed through a medical device industry forum for details on their use of particular quality control and manufacturing standards, with an invitation to participate in further contact at either an interview or focus group discussions.
   (b) Healthy (control) participants are required for a research project on musculoskeletal imaging. Consider the risks where both magnetic resonance imaging (MRI) and X-ray Computer Tomography (CT) are utilised in the imaging protocol. Note: MRI utilises very strong magnetic fields and CT ionising radiation.

4. Given an EEG data set with multiple measurements from seven participants, which sub-sampling methodology would you recommend to estimate the true error rate?
5. With relation to the diagnosis of a patient with cancer, define the following terms:

   (a) A true positive and a true negative
   (b) In this case of cancer diagnosis, explain which is likely to be the more costly or serious mistake for a computer aided diagnosis (CAD) system to make: a false positive or a false negative?

References

Bradley AP (1997) The use of the area under the ROC curve in the evaluation of machine learning algorithms. Pattern Recognit 30(7):1145–1159

Bradley AP, Longstaff ID (2004) Sample size estimation using the receiver operating characteristic curve. In: International Conference on Pattern Recognition, Cambridge, vol. 4, pp. 428–431

Breiman L, Friedman J, Olshen R, Stone C (1984) Classification and regression trees. Wadsworth, Belmont, CA

Dietterich TG, Lathrop RH, Lozano-Pérez T (1997) Solving the multiple instance problem with axis-parallel rectangles. Artif Intell 89(1–2):31–71

Duda RO, Hart PE, Stork DG (2001) Pattern classification. Wiley, New York, NY

Efron B (1982) The jackknife, the bootstrap, and other resampling plans. Society for Industrial and Applied Mathematics (SIAM), Philadelphia, PA

Fawcett T (2006) An introduction to ROC analysis. Pattern Recognit Lett 27:861–874

Friedman JH. (1995) Introduction to computational learning and statistical prediction. In: Tutorial at the International Conference on Machine Learning, Lake Tahoe, CA

Fukunaga K (1990) Introduction to statistical pattern recognition, 2nd edn. New York, NY, Academic

Hand DJ (1981) Discrimination and classification. Wiley, Chichester

Hanley JA, McNeil BJ (1982) The meaning and use of the area under a receiver operating characteristic (ROC) curve. Radiology 143:29–36

Hanley JA, McNeil BJ (1983) A method of comparing the areas under receiver operating characteristic curves derived from the same cases. Radiology 148:601–611

Landgrebe TCW, Duin RPW (2008) Efficient multiclass ROC approximation by decomposition via confusion matrix perturbation analysis. IEEE Trans Pattern Anal Mach Intell 30(5): 810–822

Landgrebe TCW, Paclik P, Duin RPW, Bradley AP (2006) Precision-recall operating characteristic curves in imprecise environments. In: International Conference on Pattern Recognition, Hong Kong, vol. 4, pp. 123–127

McLachlan GJ (1992) Discriminant analysis and statistical pattern recognition. Wiley, New York, NY

Old F (1993) Inventions, patents, brands and designs. Patent Press, Sydney

Preston N (1996) Understanding ethics. Federation Press, Sydney

Seltzer W (2005) The promise and pitfalls of data mining: ethical issues. In Proceedings of the American Statistical Association, Section on Government Statistics, Alexandria, VA: American Statistical Association, pp. 1441–1445

Swets JA, Dawes RM, Monahan J (2000) Better Decisions Through Science. Scientific American, pp. 82–87

Swets JA, Pickets RM (1982) Evaluation of Diagnostic Systems: Methods from Signal Detection Theory, Academic Press, New York

Vandewalle P, Kovacevic J, Vetterli M (2009) Reproducible research in signal processing. IEEE Signal Process Mag 26(3):37–47

Weiss S, Kulikowski C (1991) Computer systems that learn: classification and prediction methods from statistics, neural networks, machine learning, and expert systems. Morgan Kaufmann, San Mateo, CA

Zar JH (1998) Biostatistical analysis, 4th edn. Prentice-Hall, Upper Saddle River, NJ

Chapter 6
Whistle-Blowing: An Option or an Obligation?

Jong Yong Abdiel Foo

Keywords Misconduct case • Whistle-blowing • Ethical dilemma • Organisational view • Unethical practice • Unpunished act • Fraudulent data • Publication ethics • Inherent value system

Piecing It Together

The preceding chapters have covered some of the ever-growing work spectrum a biomedical engineering professional can be expected to involve with. Nevertheless, this would be sufficient to describe the significant as well as the encompassing role of the profession in the biomedical and life sciences field. From this perspective, it is then essential that not only for a biomedical engineering professional to have an inherent sound ethical value system but also for the organisation which employs the former. Known misconduct cases like that of the bioscience researcher, Hwang Woo Suk (Kakuk 2009) and the physical researcher, Jan Hendrik Schon (Reich 2009) have shown to have jeopardised any organisations associated with that individual. In addition, with the advances of media technology and the accessibility as well as popularity of social media platforms, the general public is becoming more aware of such incidences at a much quicker pace. Inevitably, the way in which misconduct is policed and corrected will very much reflect the integrity of the organisation and possibly the governmental regulatory bodies as a whole. Thus, the time frame between a misconduct case occurring and the case being known to the general community is diminishing, especially for the prominent ones. Essentially, the expectation for an organisation and/or the regulatory bodies to highlight a misconduct case

J.Y.A. Foo (✉)
Electronic and Computer Engineering Division, School of Engineering,
Ngee Ann Polytechnic, 535 Clementi Road, Singapore, Singapore 599489
e-mail: faj2@np.edu.sg

J.Y.A. Foo et al., *Ethics for Biomedical Engineers*,
DOI 10.1007/978-1-4614-6913-1_6, © Springer Science+Business Media New York 2013

or even a suspected case has become more apparent and at times, can be rather pressing. In fact, organisations and/or regulatory bodies should clarify any misconduct case at the soonest interval as the details provided by any unofficial source can sometimes induce negative yet unfounded elements to the incidence.

Over the years, the topic on ethical practices has been gaining more relevancies in both the scientific and engineering fields through education and probably deterrence. However, it may still be unclear how well professionals in the biomedical field are adopting this moral obligation. Simply, the question of whether the number of reported misconduct cases is a true reflection of all existing cases seems to be delusive (Titus et al. 2008). On the first thought, it may be easier to point the finger towards the regulatory bodies to mandate this mammoth, endless and unrewarding task. Realistically, no regulatory body can hope to be in the knowledge of all misconduct cases, but it should pave the way to install a strong ethical value culture across the country. The primary deterrent should then rest at the next tier where organisations should be the ones to establish a holistic working environment for ethical practices. Organisations should take the initiatives to promote safeguards for individuals who dare to report misconduct; this is also commonly known as whistle-blowing. Furthermore, organisations should also establish a zero tolerance for neither those who commit misconduct nor for those who turn a blind eye to it. Pivoting on these measures, it may deter potential offenders from committing the act. While the responsibility of infiltrating ethical practices seems to fall on the macro-level of society, it is also important to recognise that the moral value system at the micro-level is just as critical. Perhaps, before the efforts instituted at a macro-level can be effective, the key may lie in understanding the challenges an individual may face when attempting to report a misconduct episode or even wanting to clarify one. Simply, the decision to do what is morally right may just be dependent on addressing these challenges and providing assurance against possible or perceived repercussions.

The Dilemmas

Blowing the Whistle

Before processing further with understanding of the possible dilemmas that one may face, it is imperative to define who a whistle-blower is. It is acknowledged that whilst the following may not be an all-embracing definition of a whistle-blower, but it should entail the essences of it. Whistle-blower can be defined as an observer (any person including an employee) who witnesses an incidence where irregular practices have been adopted and reports this to the relevant authorities within the organisation voluntarily, with an unbiased viewpoint of the incidence, and without a hidden agenda.

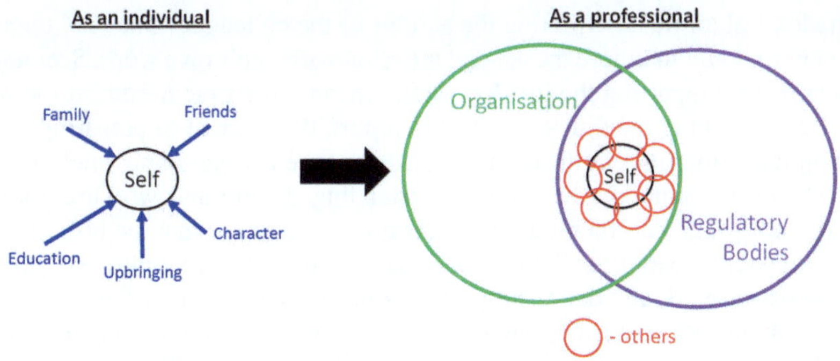

Fig. 6.1 The decision-making process of an individual is more straightforward when one needs to consider oneself only. Simply, the consideration and implication of the decision on others is generally limited. However, as a biomedical engineering professional, this process can become more complex, especially for ethical related issues. Besides the moral obligation to the general community and regulatory bodies, one needs to consider the implications of reporting in terms of the organisation, colleagues and peers

Understanding the Whys

The rationale and instinct to do what is morally right should be very much imprinted in every individual. It should also be noted that such moral behaviour is not limited only to the biomedical field. The decision to act ethically may be complicated when one interacts with others in the corporate world as illustrated in Fig. 6.1. However, before one starts finger pointing and/or condemns the unethical practices of others, it would be beneficial to understand the possible dilemmas one may have in the face of making decision that affects not only oneself but also the lives of others. In no way, should the misinterpretation that one can adopt unethical practices if one has reasonably sufficient basis or support for that heinous course of actions.

Broadly, there are two main categories of dilemma a potential whistle-blower can face. The first being the intrinsic thinking process and character of the individual while the second is one's perception of the extrinsic factors and influences. On one hand, it may be difficult to attribute an incidence to either of the categories explicitly. However, it would be useful to recognise that with each dilemma discussed herein, the confounding association between the two categories can better be manifested. However, it is recognised that there can be other perplexing considerations an individual may need to manage before making the decision to blow the whistle or not. On an individual level, it is understandable that there are many reasons for not wanting to report either a suspected or sighted incidence of unethical practices. Some of these reasons may be common feelings that everyone encounters in their daily lives. One of them may be that no one likes to accuse or be accused falsely. The thought of witnessing a colleague (even those are merely acquaintances) misbehaving can be rather daunting. There can be questions that may occur in one's mind like whether the colleague is under instructions to perform the unorthodox task or misinterpreting the actions of the colleague. One may then give the benefit of doubt to ignore the act and move on with one's own work. Second, one may fear that by reporting the incident would mean more time needed at the workplace. Besides taking the time and effort to report, the thought of preparing for and attending the numerous clarification meetings may be enough to put one off. Worst of all, should the outcome be a misunderstanding, facing and working with that particular colleague can then be rather uneasy. Third, there may be also concerns and fears about possible retaliation. One may imagine paying additional attention with one's own work and distrusting other colleagues to assist with work for fears of being sabotaged. Therefore, one may not see any value in reporting and choose to relief oneself from that possible stress. Fourth, one may assume someone else will or should report it. This attitude may be due to the lack of understanding on the significance of reporting and one's ownership at workplace. Moreover, a new staff may assume that a more senior staff should be able to detect any abnormal behaviour eventually and the senior staff would then make the report. Fifth, one may feel sympathy towards the offending colleague; especially those livelihoods are dependent on the job. One may then downplay the impact of the misbehaviour and may think that the damage can be sorted out without a career-damaging investigation, should one be warranted later. Last, one needs to have the assurance that reporting necessitates the confidence from the organisation that the incident would be examined carefully and thoroughly. This is highly dependent on the workplace culture and environment developed by the management of each organisation.

Organisational View of Whistle-Blowing

In any situation, a single observer cannot be expected to have been exposed to all instances of unethical practices. The culture within the organisation then becomes critical. Not only does the organisation need to vocalise their beliefs in ethical practices but also demonstrate their continual commitment. Particularly, the past dealings of the organisation towards offenders and whistle-blowers would very much determine the responses of employees towards future incidents of misbehaviours. However, it is postulated that the management of some organisations may also have their concerns about handling unethical practices within the organisation.

The following are some of the possible concerns each organisation may face. First, an organisation needs to consider its public image. The period between public's first knowledge of the incident and the outcome of the investigation can take its toll on the organisation. Even for incidents that turn out to be false alarms, the negative impact on the organisation's reputation can still be tedious. Thus, an organisation may try to minimise any reporting and keep any unfavourable information from reaching the press and the general public. Second, an organisation may choose an easy way out, which is to simply dismiss the accused person or even the whistle-blower, instead of conducting a thorough investigation. Besides the aforementioned concern, such organisations also hope that the problem would go away without the need for any investigation. Third, any investigations (regardless of the period) can be costly to the organisation in terms of time and money. The organisation would not only lose revenues to its competitors during the investigation but also possibly lingering intangible effects like poorer public trust and lesser business ventures. Fourth, an organisation may choose to ignore or even minimise allegations of unethical practices to especially protect key personnel within the organisation. These personnel may have substantial stakes in the organisation and/or are the ones generating revenues for the organisation. Fifth, the management of an organisation may not recognise the importance of dealing with unethical practices. This may be due to their lack of knowledge and/or being poorly equipped to investigate misbehaviour in an appropriate manner. Thus, it becomes critical that the management of any organisation is kept abreast of the development of ethical practices within their businesses. Lastly and probably most importantly, an organisation may fail to recognise its role to foster a culture of integrity. If an organisation does not institute policies that explicitly spell out its stand on unethical practices and obligate its employees to report misbehaviour, it would almost be impossible for a cohesive and coherent workplace environment to be cultured. Therefore, it must be recognised that organisational effort and commitment are significant.

Protecting the Whistle-Blowers

It has been discussed previously that in order to inculcate a culture of integrity in any workplace, it requires at least a bilateral effort. Particularly, organisations have a major role in promoting a cohesive and coherent environment. In this regard, their efforts need to be substantial. Organisation needs to delegate training programmes to educate and inculcate the value system it believes in. Elements of these training programmes should be incorporated as part of the orientation for new staffs as well as annual refresher course for existing staffs. The whole motivation for a continual effort is to remind all staffs whether new or existing, the value of ethical practices and the implications to the organisation and society. Other efforts can include having posters of its ethical values in prominent locations of the organisation and on the corporate Web site. With such an approach, this can not only create better awareness of ethical values within the organisation but also is an encouragement for staffs to query should they witness irregular practices. This can also indirectly deter potential offenders from committing the mischievous act.

Having said all that, the key consideration for an organisation to be successful in making ethical practices a norm will lie in the commitment and example setting from its management team. Only when the staffs are convinced that the management is serious, fair and appreciative of whistle-blowers, would there be a culture of open clarification. The expectation is then that the management needs to be the ones to first demonstrate its philosophical beliefs in this aspect. In addition, the management also needs to keep abreast of new regulatory requirements and explain how these can affect the organisation. Staffs may also study how the management handle preceding cases of unethical practices before the staffs can conclude the genuineness of the management in up-keeping its own preaching. Thus, it will always be easier to deter staffs from blowing the whistle than doing the otherwise correct. Anonymity is another key consideration that will determine whether the staff will step forward. The management may need to provide the necessary protection and confidence to the staffs that strict anonymity would be exercised throughout the investigation and regardless of the outcome of the incidence. In wanting staffs to whistle-blow appropriately, the role of the management cannot be undermined or shirked. In a study conducted previously (Titus et al. 2008), more than 65 % of whistle-blowers have faced at least one negative outcome as a direct result of their actions. Thus, well-defined policies and commitment to protect whistle-blowers at the organisational level then become extremely vital.

Usefulness of Independent Bodies

Besides the inherent character of oneself and the ethical culture of an organisation, the presence of an independent body, whether governmental or international, would be a useful mechanism to have within a country. The rationale of having an independent body can include potential whistle-blower(s) may be more open to verify what they have witnessed since the independent body does not report to the organisation. It can also provide an avenue where relevant information can be gathered, and it being used as a platform to facilitate regular discussions and online forums. The Committee on Publication Ethics is used herein to describe the possible roles and functions the independent body can have (Foo and Wilson 2012). It is acknowledged that the scope of the Committee on Publication Ethics may be slightly different but it is the functions of the Committee on Publication Ethics that is worth the study. In this section, the Committee on Publication Ethics will be used extensively to explain the need, the role, and the benefits of having an independent body for best ethical practices in the biomedical field of a country. To begin with, it is important to have some understandings about the Committee on Publication Ethics before relevant applications can be drawn.

The Committee on Publication Ethics is a non-profit organisation whose mission is to define best practice in the ethics of scholarly publishing. It is run by a governing council whose members are trustees of a charity and has an independent Ombudsman to adjudicate disputes between the Committee on Publication Ethics members or between them and others. The Committee on Publication Ethics provides advice to editors and publishers on all aspects of publication ethics and, in particular, how to handle cases of research and publication misconduct. It also provides a forum for its members to discuss individual cases. The Committee on Publication Ethics does not investigate individual cases but encourages editors to ensure that cases are investigated by the appropriate authorities. The membership of the Committee on Publication Ethics is open to publishers and editors of academic journals while individuals who are interested in publication ethics may also join to become its associate members. All the Committee on Publication Ethics members are expected to follow their stipulated code of conduct and the Committee on Publication Ethics would investigate complaints if any of their members are found not to have followed the code (Committee on Publication Ethics 2011). In the same manner for an independent body, organisations and their employees or interested individuals can be members and associate members of the independent body for ethical practices, respectively.

The Committee on Publication Ethics was established in 1997 with the growing concerns on how research misconduct can impact not only oneself but also the associated institutions and the country. To date, journals from several major publishers including Elsevier, Wiley-Blackwell, Springer, Taylor & Francis, Palgrave Macmillan and Wolters Kluwer have become the Committee on Publication Ethics members (Committee on Publication Ethics 2011). Similarly, group editors such as those from the Surgery Journal Editors Group have collectively agreed to adopt the guidelines established by the Committee on Publication Ethics (Committee on Publication Ethics 2010). As mentioned, the primary means by which the Committee on Publication Ethics offers support to its members is through the quarterly forum meetings where troubling cases can be presented (Wager 2010). Since its inception, the Committee on Publication Ethics has managed at least 300 cases and has posted the reports of these cases on its Web site. To assist journal editors and publishers with ethical matters, the Committee on Publication Ethics has also developed guidelines and flow charts (Committee on Publication Ethics 2011). In particular, the Committee on Publication Ethics guidelines represent a means of addressing a variety of publication ethical concerns that are becoming more prevalent such as duplicate publications and authorship misconduct issues (Committee on Publication Ethics 2010). Since these guidelines are made readily available, a guideline on publication retractions has also been widely adopted (Wager et al. 2009a, b, c). As for the flowcharts, no precise definition of the types of misconduct including plagiarism or redundant publication is provided, but it offers a decision matrix for journal editors to consider what constitutes unacceptable behaviour on a case-by-case basis (Wager 2010). Likewise, the independent body can institute guidelines and flow charts to guide organisations and employees on ethical matters. Figure 6.2 shows a proposed flow chart which the independent body can develop for organisations and potential whistle-blower to understand the process it adopts.

With the inception of the Committee on Publication Ethics, editors and publishers alike can have a better source of help and reference. Moreover, the Committee on Publication Ethics has been presenting the findings and outcomes for all the ethical cases managed onto their official Web site. With a healthy number of the established publishers and journals being the members of the Committee on Publication Ethics, as well as with a greater awareness of the Committee on Publication Ethics amongst researchers, the posting of such cases on the Committee on Publication Ethics' Web site has served as a stern warning to potential offenders. The fact that there are still on-going ethical issues occurring does imply that it is essential for the continuation of the Committee on Publication Ethics, and that there are possible gaps to be plugged. However, editors and publishers will need to continuously work

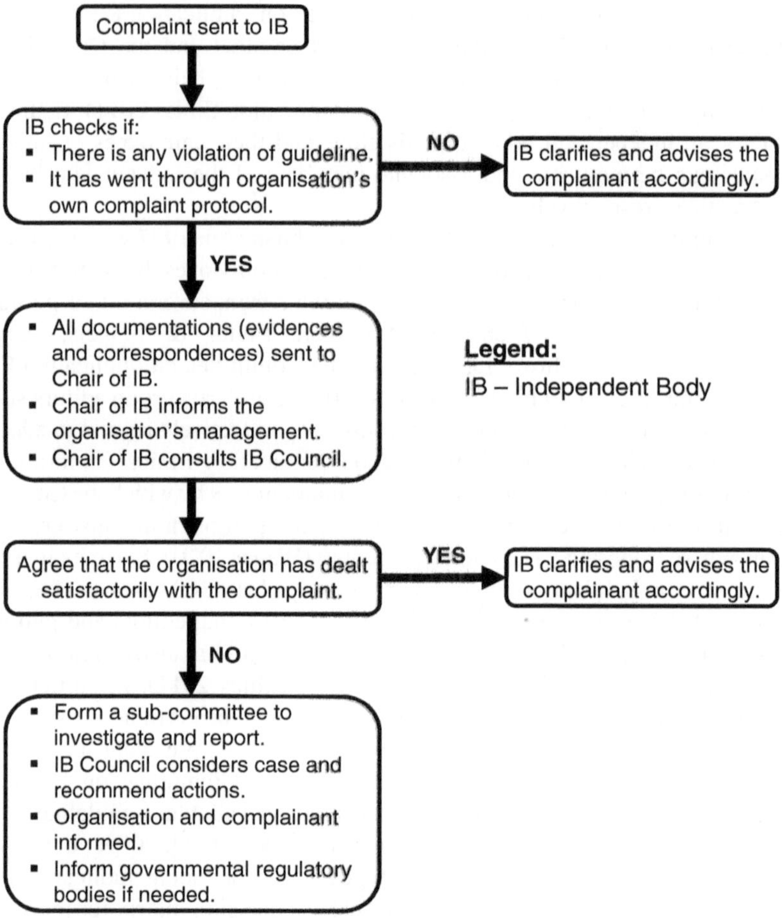

Fig. 6.2 A flow chart that an independent body can assist in the handling of complaints against an organisation that does not manage unethical practices appropriately

closely together with the Committee on Publication Ethics in order to tighten these gaps. Similarly, the independent body can provide such functions but its success will be anchored on the buy-in from all parties, that is governmental, organisational and employees.

A realistic challenge that the Committee on Publication Ethics faces is that editors may be reluctant to implement retractions when faced with threats of legal actions from authors who disagree with a retraction or whose request to retract a publication is refused; for example their name appearing on a publication without their knowledge (Wager 2010; Marusic and Marusic 2007). For the editors, they are advised to ensure that their journal has provided the information to the authors who describes the retraction procedures and explains the circumstances under which their article may be retracted prior to their manuscript submission. Thus, concerns over litigation from the editors in retracting publications can then be minimised (Sox and Rennie 2006; Wager et al. 2009d; Atlas 2004). Moreover, the information provided should also be incorporated into the publishing agreement and brought to the authors' attention again at the point when the manuscript was submitted. When possible, the choice of wordings for a notice of retraction and the reasons for retraction (to identify honest error from misconduct) should be negotiated with authors so that it is clear and informative to readers and acceptable to all parties involved (Wager et al. 2009d). The Committee on Publication Ethics can provide assistance in providing standardised conditions and circumstances where editors can retract publications. In the same way, the independent body would need to work closely with the appropriate governmental agencies to craft out guidelines for organisations about their commitment to ethical practices and the safeguarding of whistle-blowers. The independent body can then be empowered to audit organisational compliance to these guidelines. In events of non-compliance, the cause(s) should be clarified. Moreover, the independent body should protect the anonymity of whistle-blower until deemed necessary. By doing so, this can provide assurance to employees who are concerned about the implications of blowing the whistle. There are also other areas where the independent body can provide their services. First, it can incorporate on its Web site the verified offenders and/or organisations with the committed act to discourage any potential misbehaviour. With the providence of such a constituted source, it is easier to do a comprehensive check for individuals and organisations who had committed any unethical practice. Second, the independent body can serve as a resource point which tracks individuals who falsely whistle-blew and organisations that failed to comply.

Unethical practices are offensive and disturbing to not only the organisation and/or the country but also erode the trust of the general public. The formation of an independent body can create better awareness, having conceited information as well as safeguarding whistle-blowers. The purpose should be to ensure the correctness, quality and integrity of professional behaviour rather than a mean of punishing individuals and/or organisations who have misbehaved. For this to work, a tripartite effort is needed to inculcate good ethical values in younger professionals while educating the existing professionals. Once this can come to past, then the heartbeat of every professional (especially in the biomedical field) can focus on improving the well-being of patients and the community as a whole.

When Misbehaviours Are Not Addressed

In Chap. 1, the contributions from the regulatory bodies, the professional societies, the educational institutions, and the healthcare establishments towards ethical awareness have been described. In view of this, there may be misunderstandings for the whole rationale to implement greater ethical awareness and responsibility in individuals and organisations. Particularly, adopting unethical practices is one's own choice and one would have to face the thereafter consequences on one's own career. However, it must be recognised that the impact to the organisation(s) and the community can be greater at times. Thus, the importance of a whistle-blower and an organisation's culture for open clarification cannot be undermined as discussed previously. On one hand, it is acknowledged that the culprit may have suffered the consequences such as job loss and probably negative publicity for more prominent offence. It can be argued that a second chance should be given for the offender to move on. On the other hand, a critical question to ask would be by giving the offender a second chance, would there be any repercussions such as the offender recommitting a similar act. Simply, one needs to ask whether an offender has learnt the lesson and then would refrain completely from unethical practices or would the offender refuse to repent and continue the heinous behaviour as if nothing has happened before. In this section, a prior published article would be used extensively to exploit some possibilities why offenders commit the act (Foo 2011). In the same article, it also investigates the behaviour trend of offenders when their acts are exposed. All the bibliometric data used in the article have been obtained from the PubMed database (United States National Library of Medicine, Bethesda, MD).

The said article is a study that investigates the rationale of researchers publishing in academic journals with fraudulent data. The article suggests that the motivation has seemed to move from sharing scientific findings and results with peers to becoming a mere key performance indicator. The key driver in such a paradigm shift is the notion to "publish or perish" faced by researchers. Unfortunately, some cave in to the mounting pressure and misbehave in order to attain job security or recognition (Smith 2005; Sox and Rennie 2006). This can be regarded as a similar situation when professionals are in a need to deliver certain work expectations. In the article, the investigation focuses on the trend of academic authors misbehaving in the country of Singapore. Similar to other developed countries, Singapore which is one of the four Asian Tigers, dedicates a large percentage of the nation's gross domestic product into research (Kleinert 2010a, b). With the formation of public sector-based research governances such as the National Research Foundation, Singapore poises to consolidate and escalate the research and development activities within the island nation. Furthermore, strong support from the Singapore's government to enforce ethical research practices has been garn at the second World Conference on Research Integrity in the year 2010. It is also imperative to note that a high-profile episode involving scientific misconduct can implicate and tarnish the entire nation's reputation akin to how Hwang Woo-Suk had done to his home country (Aschwanden 2007; Kakuk 2009). Thus, with much high stakes involved, it would be interesting to assess the behaviour trend of offenders in publishing their works after an official retraction was made on their tainted publication. Assuming all things being equal, this trend can be extrapolated to the behaviour of offending professionals recommitting the mischievous act.

The article includes two pools of data that are analysed: published publications affiliated to an institution based in Singapore and tainted publications from the former pool (Foo 2011). The data cover publications starting from the year with the first scientific misconduct offence to the year 2010. Particularly, a total of 2,431 published publications are analysed between the year 2004 and 2010, with 9 or

Table 6.1 The figures in this table illustrate the total numbers of published publications that are affiliated to an institution in Singapore and the numbers of tainted publications according to the year

Year	2004	2005	2006	2007	2008	2009	2010	Total
Number of tainted publications	1	2	0	3	1	2	0	9
Number of published publications	278	351	388	337	345	418	314	2,431
Percentage	0.36 %	0.57 %	0.00 %	0.89 %	0.29 %	0.48 %	0.00 %	0.37 %

0.37 % being tainted publications based on the fact that there is an official retraction statement of that publication (Huang et al. 2009; Chen et al. 2009; Godge et al. 2008; Lu et al. 2007a, b; Heng 2007; Srikanth and Nagaraja 2005; Dong et al. 2005; Li and Cha 2004a, b). The number of published publications affiliated to an institution based in Singapore and the corresponding number of tainted publications are tabulated accordingly on an inter-year basis as illustrated in Table 6.1. Since the occurrence of the first publication misconduct was in the year 2004, the percentage of tainted publications to that of the total published publications has been below the 1 % mark to-date. Based on the results obtained, it can be indicative that there is no alarming trend in the number of publication misconduct in Singapore. This is despite that many academic publishers and journals have their publications made searchable online.

For Table 6.2, it shows the publication trend of authors associated with the analysed tainted publications. For the nine tainted publications, there are 29 authors associated with it. However, only 24 authors are included in the analysis as detailed in the article, since these authors are affiliated to an institution based in Singapore. Table 6.2 also illustrates the number of publications published by these disgraced authors being lower the year when the retraction of their tainted article is made known. This observation can suggest that either the offenders refrained from more heinous acts when they are exposed or their behaviour is monitored more closely by their institution. It is acknowledged that some of these authors have relocated to other countries since and their affiliation with an institution in Singapore thereby ceases. Conversely, these numbers may be an underreported number of the actual repeated offenders. It can also be observed that there was a lagging period (shaded in grey) for six of the nine tainted publications before it is eventually exposed and retracted. Unfortunately, the intangible impact and consequence to the institution and community of the lagging period cannot be easily quantified. There are also possibilities that others in the same field may have been misled unknowingly and embarked on further development based on the tainted publications.

When a nation places a large percentage of its annual gross domestic product into research, the high expectations that the returns will benefit the general population are understandable. Similar expectations may be placed on professionals in the workplace to deliver the tasks assigned to them. From the article, the proportion of tainted publications to the overall published publications may be low (<1 %). However, this figure does indicate the need for corrective actions such as educating

Table 6.2 The figures show the publication trend of authors associated with the analysed tainted publications

Year	2001	2002	2003	2004	2005	2006	2007	2008	2009	2010
Li CM*	0	0	0	0	4	3	17	15	16	22
Dong YH*		1	0	4	3	1	2	1	0	0
Zhang XF		0	0	1	0	0	0	1	0	0
Soo HM		2	0	0	0	2	0	1	1	0
Zhang LH		3	9	6	5	5	5	4	2	10
Srikanth S*		0	0	0	0	0	0	0	0	0
Lu J*				1	0	0	2	0	0	0
Wang XJ				2	1	0	1	1	0	0
Yang X				0	0	1	0	0	0	0
Ching CB				6	1	2	6	5	2	5
Heng BC*				9	25	18	16	5	3	5
Lu L*				2	4	1	0	0	0	0
Yu L				0	2	0	0	1	0	0
Kwang J				7	6	7	4	7	10	7
Godge MR*					0	0	0	1	0	0
Kumar PP					1	2	3	2	0	1
Chen Y*						0	0	0	0	0
Narayanswamy B						0	0	0	0	0
Chew M						0	0	0	0	0
Huang Q*						2	3	1	2	2
Wu YT						0	0	2	1	2
Tan HL						0	0	2	2	1
Ong CN						6	9	6	11	8
Shen HM						5	8	5	3	5

Only 24 authors are included as they are affiliated to an institution based in Singapore
The underlined number denotes the figures before and on the year the tainted publication is retracted
The grey shaded area denotes the lagging period between the tainted publication being published and it eventually being retracted
The asterisk (*) symbol denotes the first author of each tainted publication

the importance of ethical behaviour for those involved in research or having some forms of rewards system to encourage ethical practices. It is also important to recognise that the numbers in the study do not include ethical cases like authorship irregularities and unethical conducts in research experiments which are much harder to track unless someone reports it. In addition, the low figure presented herein is also coupled with the fact that there are limitations of the study (Foo 2011). If the figure can be correctly extrapolated, the actual figure for research misconduct may just be a tip of the iceberg. In addition, one incidence may be all it needs to tarnish the indirect association(s), be it the institution or country. More importantly, the more worrying question would be how likely these offenders would repeat their once-mischievous act again and with their previous experience of being exposed, would they become "smarter" in concealing their mischievous act. Likewise, this can also happen to an organisation when an employee misbehaves.

Thus, a well-coordinated long-term commitment and effort from the organisation can escalate the momentum to combat misbehaviours.

Another point to highlight is that there are evidences that "positive" results that support the experimental hypothesis against a "null" hypothesis of no effect (Fanelli 2010) are more likely to be published in notable journals (Dwan et al. 2008; Murtaugh 2002) and to be cited by peers (Etter and Stapleton 2009; Leimu and Koricheva 2005). Since publications with a "positive" result stands a better chance to be accepted, this may influence researchers to select which results to be published or to manipulate the data to orchestrate the expected outcomes (Fanelli 2010; Chan et al. 2004). This is expected as the career of researchers is usually evaluated by counting the number of publications listed in their resume, and the impact factor of the journals their articles are being published in Smith (2005) and Sox and Rennie (2006). Similarly, employees who are able to demonstrate "positive" results in their work usually receive rewards and recognition. Thus, there are incentives and temptations in this regard for employee to also fabricate results. In a study conducted in the USA (Martinson et al. 2005), it collates a survey of questionable behaviours that threatens the integrity of researchers. A total of 1,768 mid-career researchers and 1,479 early-career researchers provided their responses to that survey. The former group had an average of 9 years more of working experience than the latter. The study reports that 38 % of the mid-career researchers have engaged in at least one of the misconduct behaviours (as defined in the study) during a period of 3 years prior to the study while for the early-career researchers, the figure is only 28 %. The results can imply that the norms in the working environment can constitute to a learnt behaviour of an individual. Thus, when an offender is left unpunished, it is likely that the offender may repeat the act as well as incite others to follow suit. It is believed that this trend can also be applicable to working professionals with all things being held equal.

There are definitely gaps to be plugged in ensuring that a cohesive and conducive work environment is maintained. Besides personal convictions, it is also important to recognise there can be adverse effects of not addressing misbehaviours first at the organisational level, and then eventually at the national level. Particularly, offenders when left unpunished would continue with their mischievous acts based on the postulated results in the article (Foo 2011) and the study (Martinson et al. 2005). The results may also suggest that it may become a negative influence to others in their commitment to ethical practices. Moreover, the community would lose confidence in any organisation or government which deals with such offenders inappropriately. Therefore, there is a need for stricter disciplinary actions against offenders and it being policed religiously at all levels of society.

Ethical Practices as a Way of Life

Working in proximity to human lives has increased the relevance of ethics to any biomedical engineering professional. Particularly, the various possibilities and scope of how a biomedical engineering professional can intervene have been described in the preceding chapters; namely, in clinical settings, in designing and developing medical instrumentations and devices as well as when analysing the data acquired from human subjects. Simply, a biomedical engineering profession is not merely performing a technical job, but there are broader obligations to the society. While discharging their duties, biomedical engineering professionals must have in mind the prior need of protecting lives and enhancing the well-being of patients. It is recognised that the coverage of this book may not be all encompassing to the ever expanding spectrum of tasks biomedical engineering professionals are involved with. However, it is hoped that through this book, each and future biomedical engineering professional can acquire the necessary tools to recognise and approach ethical issues holistically. However, there should be the understanding that application of these tools may often not reach any consensus, even amongst other healthcare workers including fellow biomedical engineering professionals. As mentioned previously, the critical thing is for one to understand the context of ethics first to oneself, then how this is applicable in a specific circumstance and how that circumstance should not affect one's decision. It is also worth noting that from the viewpoint of a scientific endeavour, any data analysis illustrated in this book is not always precisely reproducible because the data is not static in that further relevant information may be made available later or the authenticity of present information may be discounted in the future. This can imply that the data presented herein may vary with any studies performed after each passing period. However, the understanding, methodologies and analyses should still remain the same as well as it being statistically valid. Hence, it is anticipated that the contributions of this book can provide a framework from which future endeavours involving ethical practices may be able to draw useful insights and examples from within.

Case Study Examples

The following are three known cases that were involved with unethical practices. These cases are interesting to study because the individual(s) connected to each case were fully aware of the implications and consequences of their course of actions; however, they still opted to act otherwise. From the prognosis to the aftermath of each case, it enfolds the various ethical principles involved and the rationale pegged to their unfortunate decision.

1. The case involving the bioscience researcher, Dr. Hwang Woo Suk.

Suggested further reading

Kakuk P (2009) The legacy of the Hwang case: research misconduct in biosciences. Sci Eng Ethics 15:545–562.

Background

This article focuses on an infamous case of Dr. Hwang Woo Suk. He was a former South Korean professor of theriogenology and biotechnology at the Seoul National University (SNU). Until the year 2006, Hwang was considered to be an outstanding and pioneering expert in stem cell research and was regarded as a national hero in South Korea. However, the allegations of fraud and research misconduct against him were soon confirmed. In the stem cell research field, he was best known for two articles published in a renowned academic journal, Science, where he fraudulently reported to have succeeded in creating human embryonic stem cells by cloning. These articles have been retracted from publication by the journal editorial after evidences of large amount of fabricated data were found. Thus, it is interesting to review the circumstances leading to the decision by Hwang to fabricate data in his stem cell research as well as the possible ethical implications and lessons that can be derived from this incident.

2. The case involving the physical researcher, Jan Hendrik Schon.

Suggested further reading

Reich ES (2009) The rise and fall of a physics fraudster. Phys World 22:24–29.

Background

This article centres on the fall of a rising star in the fields of physics, materials science and nanotechnology. Jan Hendrik Schon was a former young physicist at the Bell Labs in the USA. He was emerging with remarkable speed for his fraud research work on an evolutional ability to transform the properties of materials by the application of an electric field. His fabricated discoveries had misled many scientists and dozens of laboratories where much time and money were wasted. Schon managed to coax the management of top research institutions in the USA to back and promote his claims as well as renowned international academic journals to publish his fraudulent results. It is believed that Schon was doing science backwards where he started from the conclusion he wanted and then assembled data to show it. Therefore, it would be important to examine the motivations of Schon to commit the heinous acts as well as the likely ethical complications and lessons that can be learnt from this incident.

3. The case involving the academic journal, Folia Phoniatrica et Logopaedica.

Suggested further reading

Foo JYA (2011) Impact of excessive journal self-citations: a case study on the Folia Phoniatrica et Logopaedica Journal. Sci Eng Ethics 17:65–73.

Background

This article evolves from a decision by the editors of the Swiss journal, Folia Phoniatrica et Logopaedica. Particularly, it is concerning the journal impact factor (JIF) value of their said journal. The JIF value is the most widely used measure by academia to determine the scientific contents and importance published in a journal. For a number of years, the JIF of this Swiss journal has not seen any significant improvement (fluctuating between 0.339 and 0.655). In the year of question, the JIF value of the journal registered a remarkable JIF increment (of 119 %) to 1.439. It is believed that the journal can achieve such a prominent JIF improvement by publishing a single editorial article that self-cited 66 of its own articles published. However, the journal has been revoked of any JIF value in the following year. Hence, it is worth to analyse the rationale of the editors in deciding to publish such an article as well as the ethical implications and lessons associated with this incident.

References

Aschwanden C (2007) Seeking an international dialogue on research integrity. Cell 131:9–11

Atlas MC (2004) Retraction policies of high-impact biomedical journals. J Med Libr Assoc 9:242–250

Chan AW, Hrobjartsson A, Haahr MT, Gotzsche PC, Altman DG (2004) Empirical evidence for selective reporting of outcomes in randomized trials – comparison of protocols to published articles. JAMA 291:2457–2465

Chen Y, Narayanswamy B, Chew M (2009) Polyurethane hydrogel colored contact lens with gradient water content. J Biomed Mater Res B Appl Biomater 89:576, Retraction in: J Biomed Mater Res B: Appl Biomater 89:576

Committee on Publication Ethics (2010) Consensus statement on the adoption of the COPE guidelines. Ann Surg 252:2

Committee on Publication Ethics (2011) About COPE [Internet] [Cited 12 Dec 2011]. Available from: http://publicationethics.org/about

Dong YH, Zhang XF, Soo HM, Greenberg EP, Zhang LH (2005) The two-component response regulator PprB modulates quorum-sensing signal production and global gene expression in Pseudomonas aeruginosa. Mol Microbiol 56:1287–1301, Retraction in: Pugsley A (2008) Mol Microbiol 69:780

Dwan K, Altman DG, Arnaiz JA, Bloom J, Chan AW, Cronin E, Decullier E, Easterbrook PJ, Von Elm E, Gamble C, Ghersi D, Ioannidis JP, Simes J, Williamson PR (2008) Systematic review of the empirical evidence of study publication bias and outcome reporting bias. PLoS One 3:e3081

Etter JF, Stapleton J (2009) Citations to trials of nicotine replacement therapy were biased toward positive results and high-impact-factor journals. J Clin Epidemiol 2009(62):831–837

Fanelli D (2010) Do pressures to publish increase scientists' bias? An empirical support from US States data. PLoS One 5:e10271

Foo JYA (2011) Trend of tainted publications and their authors' publication profile: a review of Singapore based on PubMed database. Int J Ethics 7(2):article 3.

Foo JY, Wilson SJ (2012) An analysis on the research ethics cases managed by the Committee on Publication Ethics (COPE) between 1997 and 2010. Sci Eng Ethics 18:621–631. doi:10.1007/s11948-011-9273-3

Godge MR, Purkayastha A, Dasgupta I, Kumar PP (2008) Virus-induced gene silencing for functional analysis of selected genes. Plant Cell Rep 27:209–219, Retraction in: Plant Cell Rep (2009) 28:335

Heng BC (2007) Is it ethically justifiable to cryopreserve oocytes and ovarian tissues of healthy women not facing premature ovarian failure, but seeking to extend their biological clocks? Hum Fertil 10:49–50, Retraction in: Hum Fertil 10:193
Huang Q, Wu YT, Tan HL, Ong CN, Shen HM (2009) A novel function of poly(ADP-ribose) polymerase-1 in modulation of autophagy and necrosis under oxidative stress. Cell Death Differ 16:264–277, Retraction in: Cell Death Differ (2010) 17:1944
Kakuk P (2009) The legacy of the Hwang case: research misconduct in biosciences. Sci Eng Ethics 15:545–562
Kleinert S (2010a) Singapore embraces international research integrity. Lancet 376:400–401
Kleinert S (2010b) Singapore statement: a global agreement on responsible research conduct. Lancet 376:1125–1127
Leimu R, Koricheva J (2005) What determines the citation frequency of ecological papers? Trends Ecol Evol 20:28–32
Li CM, Cha CS (2004a) Porous carbon composite/enzyme glucose microsensor. Front Biosci 9:3479–3485
Li CM, Cha CS (2004b) Porous carbon composite/enzyme glucose microsensor. Front Biosci 9:3324–3330
Lu L, Yu L, Kwang J (2007a) Baculovirus surface-displayed hemagglutinin of H5N1 influenza virus sustains its authentic cleavage, hemagglutination activity, and antigenicity. Biochem Biophys Res Commun 358:404–409, Retraction in: Biochem Biophys Res Commun (2009) 388:620
Lu J, Wang XJ, Yang X, Ching CB (2007b) Characterization and selective crystallization of famotidine polymorphs. J Pharm Sci 96:2457–2468, Retraction in: Lu J, Wang XJ, Yang X, Ching CB (2008) J Pharm Sci 97:1629
Martinson BC, Anderson MS, de Vries R (2005) Scientists behaving badly. Nature 435:737–738
Marusic M, Marusic A (2007) Threats to the integrity of the Croatian Medical Journal. Croat Med J 48:779–785
Murtaugh PA (2002) Journal quality, effect size, and publication bias in metaanalysis. Ecology 2002(83):1162–1166
Reich ES (2009) The rise and fall of a physics fraudster. Phys World 22:24–29
Smith R (2005) Investigating the previous studies of a fraudulent author. BMJ 331:288–291
Sox HC, Rennie D (2006) Research misconduct, retraction, and cleansing the medical literature: lessons from the Poehlman case. Ann Intern Med 144:609–613
Srikanth S, Nagaraja AV (2005) A prospective study of reversible dementias: frequency, causes, clinical profile and results of treatment. Neurol India 53:291–294, Retraction in: Neurol India (2007) 55:5
Titus SL, Wells JA, Rhoades LJ (2008) Repairing research integrity. Nature 453:980–982
Wager E (2010) The committee on publication ethics flowcharts. Chest 137:221–223
Wager E, Barbour V, Yentis S, Kleinert S (2009a) Retractions: guidance from the Committee on Publication Ethics (COPE). Croat Med J 50:532–535
Wager E, Barbour V, Yentis S, Kleinert S (2009b) Retractions: guidance from the Committee on Publication Ethics. J Crit Care 24:620–622
Wager E, Barbour V, Yentis S, Kleinert S (2009c) Retractions: guidance from the Committee on Publication Ethics (COPE). Maturitas 64:201–203
Wager E, Barbour V, Yentis S, Kleinert S (2009d) Retractions: guidance from the Committee on Publication Ethics (COPE). Front Matter 10:1–6

Index

J.Y.A. Foo et al., *Ethics for Biomedical Engineers*,
DOI 10.1007/978-1-4614-6913-1, © Springer Science+Business Media New York 2013

MIX
Papier aus verantwortungsvollen Quellen
Paper from responsible sources
FSC® C105338

If you have any concerns about our products,
you can contact us on
ProductSafety@springernature.com

In case Publisher is established outside the EU,
the EU authorized representative is:
Springer Nature Customer Service Center GmbH
Europaplatz 3, 69115 Heidelberg, Germany

Printed by Libri Plureos GmbH
in Hamburg, Germany